Geology and Mineral Resources of the Gasquet Quadrangle of California-Oregon

by US Department of Interior

with an introduction by Kerby Jackson

This work contains material that was originally published in 1953.

This publication is within the Public Domain.

This edition is reprinted for educational purposes
and in accordance with all applicable Federal Laws.

Introduction Copyright 2015 by Kerby Jackson

Introduction

It has been sixty years since the US Department of Interior released their important publication "Geology and Mineral Resources of the Gasquet Quadrangle of California-Oregon". First released in 1953, this work has been unavailable to the mining community since those days, with the exception of expensive original collector's copies and poorly produced digital editions.

It has often been said that *"gold is where you find it"*, but even beginning prospectors understand that their chances for finding something of value in the earth or in the streams of the Golden West are dramatically increased by going back to those places where gold and other minerals were once mined by our forerunners. Despite this, much of the contemporary information on local mining history that is currently available is mostly a result of mere local folklore and persistent rumors of major strikes, the details and facts of which, have long been distorted. Long gone are the old timers and with them, the days of first hand knowledge of the mines of the area and how they operated. Also long gone are most of their notes, their assay reports, their mine maps and personal scrapbooks, along with most of the surveys and reports that were performed for them by private and government geologists. Even published books such as this one are often retired to the local landfill or backyard burn pile by the descendents of those old timers and disappear at an alarming rate. Despite the fact that we live in the so-called "Information Age" where information is supposedly only the push of a button on a keyboard away, true insight into mining properties remains illusive and hard to come by, even to those of us who seek out this sort of information as if our lives depend upon it. Without this type of information readily available to the average independent miner, there is little hope that our metal mining industry will ever recover.

Though this volume may not at first seem to be of great importance to gold miners, I feel that those miners with an interest in smelting and refining their finds, especially those recovered from lodes, will find the processes outlined to be of great value.

This important volume and others like it, are being presented in their entirety again, in the hope that the average prospector will no longer stumble through the overgrown hills and the tailing strewn creeks without being well informed enough to have a chance to succeed at his ventures.

Please note that at times it is necessary to rearrange illustration plates in these texts. Any illustrations not found in their original sequence may be found following the index.

Kerby Jackson
Josephine County, Oregon
January 2015

www.goldminingbooks.com

A CONTRIBUTION TO ECONOMIC GEOLOGY

GEOLOGY AND MINERAL RESOURCES OF THE GASQUET QUADRANGLE, CALIFORNIA-OREGON

By Fred W. Cater, Jr., and Francis G. Wells

ABSTRACT

The Gasquet quadrangle is in the sparsely settled mountains of northwestern-most California. The maximum relief is nearly 5,000 feet, altitudes ranging from about 350 feet to nearly 5,800 feet. Most of the streams in this area of mild humid climate are perennial throughout their courses.

The rocks of the quadrangle may be divided into three groups: pre-Tertiary, Tertiary, and Quaternary. The pre-Tertiary rocks consist of highly deformed and in part metamorphosed sedimentary and volcanic rocks intruded by many igneous masses. The sedimentary and volcanic rocks comprise two formations, the Dothan and the Galice, both of Late Jurassic age. The Dothan formation is thought to be the older and consists of a great thickness, perhaps 15,000 feet, of graywacke, sandstone, shale, chert, and conglomerate. The Galice formation includes a lower metavolcanic rock unit at least 7,000 feet thick, consisting of andesitic flows, breccias, and tuffs, and an upper sedimentary rock unit consisting of not less than 3,000 feet of slate and phyllite with interbedded tuffaceous sandstone. The lower metavolcanic rock unit may be correlated with the Rogue formation (Wells and Walker, in press). The pre-Tertiary plutonic rocks have been divided on a basis of composition into two groups: ultramafic rocks and dioritic-gabbroic rocks. The ultramafic rocks are older and include dunite, saxonite, lherzolite, wehrlite, pyroxenite, and their alteration product, serpentine. The dioritic-gabbroic rocks range in composition from quartz diorite to gabbro, although gabbro and diorite are by far the most abundant.

A profound unconformity separates the pre-Tertiary rocks from the Tertiary rocks. Tertiary rocks crop out over relatively small areas between altitudes of 1,900 to 2,200 feet and include thin beds of soft shale, siltstone, and sandstone called here upper Miocene beds; poorly sorted stream gravels of somewhat later age are late Miocene or early Pliocene. A few albite rhyolite porphyry dikes are thought to be Tertiary.

Quaternary deposits consist of stream and terrace sand and gravel, talus debris, and landslide material.

The pre-Tertiary structural features include isoclinal folds, generally trending somewhat east of north, cut by reverse faults of the same trend and also by two systems of later cross faults, one trending northwestward and the other northeastward. Contacts between the Dothan and Galice formations have been destroyed by plutonic intrusions, so that major structural features could not be deciphered.

Post-Miocene deformation has resulted in vertical uplift and slight tilting within the quadrangle but no folding or faulting of Miocene or later rocks.

The history of the quadrangle since the Early Cretaceous has been dominantly one of erosion. By Miocene time this long period of erosion had produced the

Klamath peneplane, the remnants of which are still the dominating topographic features in the area. The Klamath cycle of erosion ended with uplift and canyon cutting followed by slight downdropping, flooding of the lower ends of valleys, and deposition of the upper Miocene beds. A number of partial erosion cycles of post-Wymer age have also been recognized, the latest of which forms a series of stream terraces 50 to 75 feet above the present channels of the major streams.

The mineral resources of the area include deposits of chromite, gold, copper, mercury, platinum, and asbestos. Of these, the value of the chromite produced far overshadows the combined value of the other mineral products.

INTRODUCTION

FIELD WORK AND ACKNOWLEDGMENTS

A considerable part of the field work on which this report is based was done in 1943 and 1944; much of the areal mapping of nonultramafic rocks was done in September and October 1946. The earlier work consisted largely of mapping peridotite and studying the economically significant features of these rocks as related to chromite deposits—a phase of the U. S. Geological Survey's war time Strategic Minerals Investigations program. Only the main features of the large peridotite masses were delimited. Somewhat more systematic traversing of these masses would have added little additional pertinent information, and closer mapping was not justified by the scale of the map or the purposes of the investigation. The mapping in 1946 was a cooperative project between the California Division of Mines and the U. S. Geological Survey. The writers were assisted during the later mapping by Ralph J. Newton. F. G. Wells contributed much to the preparation of this report, which was written largely by Cater.

Previous work in the quadrangle has been limited to general reconnaissance by Diller (1902), Hershey (1911), and Maxson (1933) and detailed studies of the chromite deposits by Wells, Cater, and Rynearson (1946). The Kerby quadrangle, adjoining the Gasquet on the north, has been mapped by Wells and others (1949), and work in the Gasquet quadrangle extended this mapping southward.

The aid and courtesy extended the writers by the people of the Gasquet quadrangle, especially Mr. Charles Bennett, as well as Messrs. Robert R. Steven and William Steven, of Big Flat, are hereby acknowledged. Messrs. H. C. Obye and C. D. Cameron, supervisor and engineer, respectively, of the Siskiyou National Forest, within which the quadrangle was then located, were unfailingly helpful.

LOCATION AND GENERAL FEATURES

The Gasquet quadrangle covers an area of about 215 square miles in the northwestern corner of California and is bounded by the meridians 123°45′ and 124°00′ W. and the parallels 41°45′ and 42°00′ N.

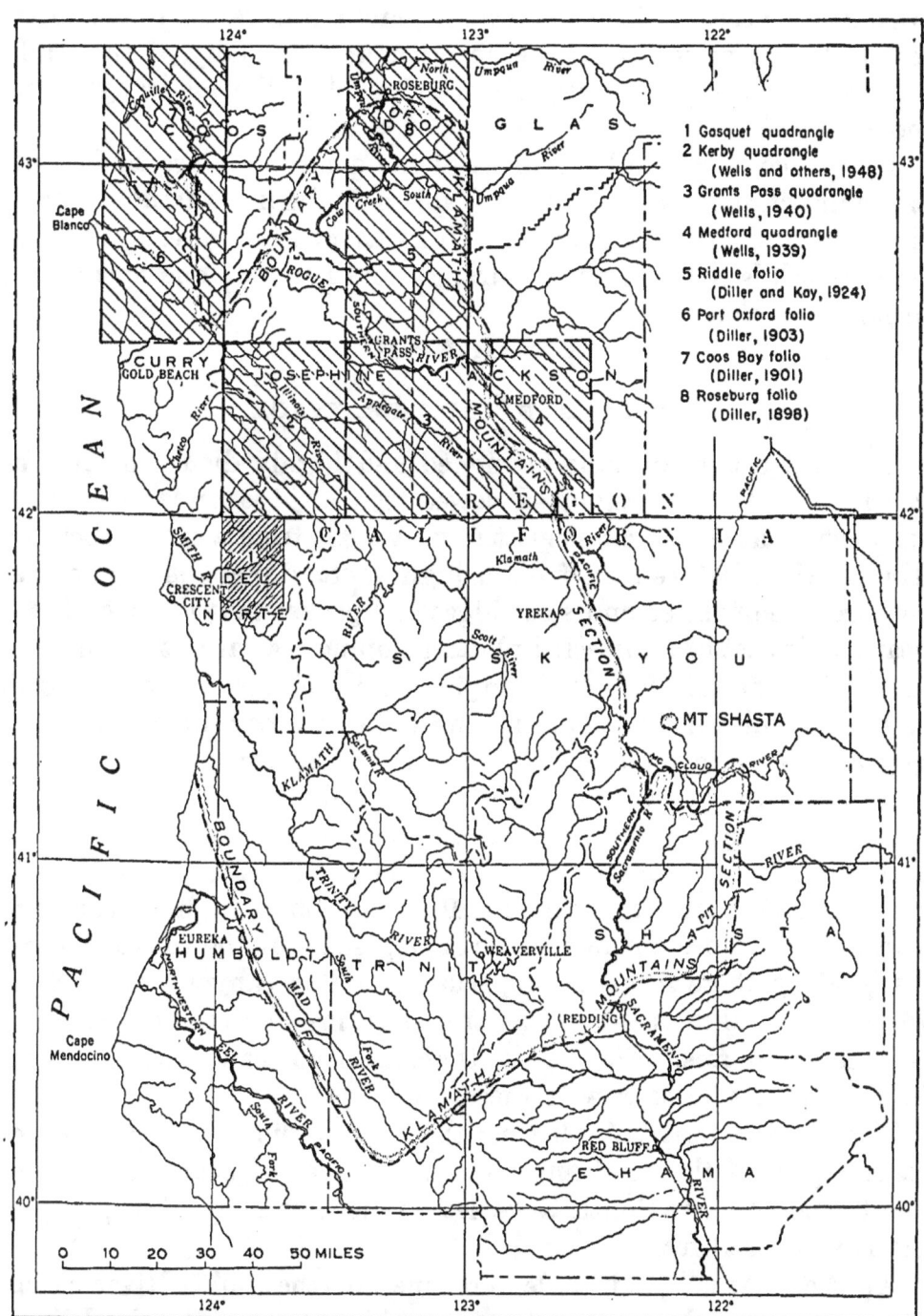

FIGURE 9. Map of northwestern California and southwestern Oregon showing the Klamath Mountains section and the location of the Gasquet quadrangle and quadrangles described in other publications.

A strip about a quarter of a mile wide along the north edge of the quadrangle is in Oregon. (See fig. 9.) The quadrangle lies along the west slope of the Klamath Mountains.

The area is sparsely populated; less than 50 people are permanent residents of the quadrangle, and nearly all of them live along the Middle Fork of Smith River. A paved highway, U. S. Highway 199, traverses the center of the quadrangle, and a number of Forest Service roads and various trails reach other areas. Many of the trails, however, are overgrown with brush, and much of the quadrangle is remote from either roads or trails.

The principal natural resources of the quadrangle are timber and chromite; chromite mining is profitable only during wartime. Vacation resorts capitalize on the beautiful scenery in the area. Very little livestock is raised in the quadrangle and less than one square mile of land is arable.

GEOGRAPHY

TOPOGRAPHY AND DRAINAGE

The Gasquet quadrangle is rugged and highly dissected, but the general aspects of the region as viewed from a ridge top is very different from what one might suspect in viewing it from a canyon bottom. The striking feature seen from a high vantage point is the general concordance of the summits of ridges and peaks; thus the overall appearance is that of a dissected plateau sloping seaward at an angle of about 2°. (See pl. 12.) This impression is heightened by the broad, rolling summits of the mountains in the western part of the quadrangle; eastward, however, most of the ridges are sharp and narrow, but the concordance of summits is maintained.

The quadrangle is cut by numerous steep-walled canyons, many of them 2,000 feet or more deep. (See pl. 11.) In areas of particularly resistant rock, nearly vertical cliffs 100 feet or more high are common. Landslides are prominent features along some of the canyons, especially along those in areas of peridotite. Box canyons as much as 75 feet deep wall in many of the larger streams, more commonly along their lower courses. The rims of these inner box canyons are fringed in many places by narrow terraces.

The relief is nearly 5,000 feet; the highest summits in the southeastern part of the quadrangle reach altitudes of nearly 5,300 feet, and the point where Smith River leaves the west side of the quadrangle is about 350 feet.

All the Gasquet quadrangle is drained by the Smith River except a small area in the northeast corner which drains into the Illinois River. The major stream, the Middle Fork of Smith River, flows through most of its course in a straight west-southwesterly direction and approximately bisects the quadrangle.

CLIMATE AND VEGETATION

The climate of the Gasquet quadrangle is mild and humid. The average annual rainfall at Crescent City, a few miles west of the quadrangle, is 74 inches, and the average of a 6-year period (1904–1910) at Monumental was 153.54 inches (U. S. Department of Agriculture, 1926). The months of July and August and much of September are usually dry. The higher mountains are usually covered with snow from November to April, and in exceptional years snow may remain above 4,500 feet until June; at lower altitudes it seldom stays long. Below an altitude of 400 feet killing frosts are rare.

The abundant rainfall favors a lush, dense vegetation except on areas underlain by peridotite which are rapidly and deeply drained; these areas are nearly devoid of soil and support only a sparse growth of yellow pine and stunted manzanita. A few scattered coast redwoods grow at altitudes up to 2,000 feet on and near French Hill but are absent elsewhere. Great virgin forests of fir, spruce, pine, and cedar cover large areas in the quadrangle. The most pervasive vegetation, however, consists of extremely dense, almost impenetrable growths of manzanita, salal, scrub oak, rhododendron, and others. This brush forms dense understories even in the heavily forested areas and is a severe hindrance to travel and mapping.

GEOLOGY

The Gasquet quadrangle has the complicated stratigraphy and structure common to the rest of the Klamath Mountains. The rocks may be considered conveniently in two categories: the pre-Tertiary rocks and those of Tertiary and of Quaternary age. The pre-Tertiary rocks consist of highly folded metamorphosed sedimentary and volcanic rocks in which are intruded numerous igneous masses. The Tertiary and Quaternary rocks consist of undeformed soft sediments and perhaps a few rhyolite dikes.

The pre-Tertiary supracrustal rocks are of Jurassic age and have been divided into two formations: the Dothan and the Galice. The Dothan formation is probably the older and is composed largely of graywacke and lesser amounts of sandstone, shale, chert, and intercalated volcanic rocks. It aggregates a great thickness, probably more than 15,000 feet, but not more than 3,000 feet of the formation is exposed in the Gasquet quadrangle. The Galice formation within the quadrangle comprises a series, probably not less than 10,000 feet thick, of meta-andesite flows and pyroclastic rocks, black slate, mudstone, and tuffaceous sandstone. Fossils are rare in both formations.

The pre-Tertiary intrusive rocks include several varieties of ultramafic rocks, gabbro, hornblende diorite, and subordinate amounts of quartz diorite, and other varieties; all are in part altered. The intrusions are probably Late Jurassic and Early Cretaceous.

The Tertiary rocks are separated from the pre-Tertiary rocks by a pronounced unconformity. Tertiary sedimentary rocks crop out over relatively small areas and include very soft shale, siltstone, and sandstone, and irregular beds of stream-deposited gravel. Fossils found in the shale indicate that it is late Miocene; the gravel is somewhat younger but may also be late Miocene. A few rhyolite porphyry dikes that cut the pre-Tertiary rocks are thought to be of Tertiary age.

The Quaternary deposits consist of stream and terrace sand and gravel, talus debris, and landslide material.

The following is a generalized section of the sequence and thickness of the supracrustal rocks in the Gasquet quadrangle:

Generalized section of the supracrustal rocks exposed in the Gasquet quadrangle

Age	Formation		Thickness (feet)	Lithology
Quaternary_____	Alluvium_____		0–20_____	Stream and terrace gravels, sand, and silt. Talus debris. Unconsolidated.
Tertiary, Miocene(?)__	Gravel_____		0–70_____	Coarse stream gravels and sand; largely unconsolidated, but loosely cemented locally.
	—Unconformity—			
Miocene (Upper)_____	Upper Miocene beds_		Probably less than 50.	Soft, friable yellow shale and silt that weather red.
	—Unconformity—			
Jurassic, Upper Jurassic.	*Galice formation*	Slate_____	At least 3,000_____	Black fissile slate and phyllite with interbedded light- to dark-gray tuffaceous sandstone. One small lens of limestone and minor intercalated volcanic rocks.
		Metavolcanic rocks.	At least 7,000_____	Altered volcanic flows and pyroclastic rocks; composition ranges from rhyolitic to basaltic but consists very largely of andesite. A few interbedded lenses of black slate becoming more abundant higher in the section.
	—Relation unknown			
	Dothan formation (sedimentary and volcanic rocks).		Perhaps 15,000_____	Hard dark-gray thick-bedded graywacke, and lesser amounts of sandstone, shale, chert, and intercalated, largely basaltic, volcanic rocks.

PRE-TERTIARY ROCKS

SUPRACRUSTAL ROCKS

DOTHAN FORMATION

Rocks of the Dothan formation underlie an area of slightly less than 1 square mile in the northwestern corner of the Gasquet quadrangle but are much more extensively exposed to the west and north. (See pl. 12.)   Areas underlain by the Dothan formation are characterized by rougher topography than those underlain by the Galice formation.   This is due to a greater heterogeneity in the lithology of the Dothan rocks.   In common with other areas underlain by non-ultramafic rocks, areas underlain by the Dothan formation are covered with timber and very dense brush.   The thickness of the formation within the Gasquet quadrangle is unknown but ranges from 3,000 to 18,000 feet.

The Dothan formation is composed largely of graywacke and lesser amounts of sandstone, shale, and intercalated volcanic rocks.   The graywackes are dark-gray hard thick-bedded rocks made up of arkosic material and fragments of other rocks, predominantly shale. As a rule bedding in the massive graywacke is very obscure. Interlayered with the graywacke are thin beds of hard sandstone containing a large proportion of quartz grains; both graywacke and sandstone are metamorphosed. Fractures pass through, rather than around, sand grains. The darker minerals in the graywacke in general have altered to chlorite, and sericite is abundant in the sandstones. Numerous irregular narrow veins of white quartz cut both rocks. Shale is much less abundant than sandstone and forms thin layers interbedded with the graywacke. The shales generally lack the slaty cleavage so characteristic of the argillaceous rocks in the overlying Galice formation, but recrystallization has advanced fully as far, forming chlorite and mica. Where the shale occurs as thin interbeds between massive graywacke or sandstone layers, it is commonly sheared. Lenses of chert are common. Intercalated with these sediments are thick layers of green volcanic agglomerate and fine-grained flows, most of which appear to be basaltic.

The lithology of the Dothan rocks exposed within the Gasquet quadrangle corresponds with the lithology of the upper parts of zone 2 and zone 3 of the formation as exposed in the Rogue River Canyon 50 miles to the north (Wells and Walker, in press). Both the sedimentary and volcanic rocks have been cut by intrusions of ultramafic and dioritic rocks.

No fossils have been found in the Dothan formation in northern California, and they are rare in exposures of this formation elsewhere. Those found have been confusing, either because the fossils were not diagnostic or because of doubt concerning the relation of fossiliferous beds to the Dothan.

The formation was considered by earlier workers in the region (Diller, 1914; Butler and Mitchell, 1916; Maxson, 1933) to be younger than the Galice, a point of view first questioned by Taliaferro (1942). The formation has now been traced by systematic mapping with but two minor breaks from the type locality on Cow Creek, Oreg., southward to the Gasquet quadrangle. Although the evidence is not yet conclusive, indications are that the Dothan is older than the Galice.

The Dothan rocks in the northwest corner of the Gasquet quadrangle plunge southward under, and are generally conformable with, the peridotite sheet of Josephine Mountain. To the south, on the Middle Fork of Smith River, the peridotite in turn underlies southward plunging metavolcanic rocks of the Galice formation that are known to be in normal position. Recent mapping by Wells and Walker (in press) in the Galice quadrangle in southwestern Oregon, on the Rogue River between Mule and Whisky Creeks, indicates the Dothan rocks are in normal sequence; they grade upward into a thick series of metavolcanic rocks which farther east are in contact with slate of the Galice formation. Considerable deformation along this contact between the metavolcanic rocks and the slate of the Galice is thought to be due to local crushing, shearing, and drag folding. Thus, on the basis of available data, it appears reasonable to assume that the Dothan underlies the Galice and, although older than the Galice, is still probably no older than Late Jurassic.

GALICE FORMATION

METAVOLCANIC ROCKS

The lower part of the Galice formation consists of altered volcanic flows and pyroclastic rocks and thin subordinate layers of slate. These rocks are exposed over wide areas in the Klamath Mountains and underlie considerable areas in the eastern half and southwestern quarter of the Gasquet quadrangle. Three areas of these rocks, separated by belts of slate and diorite, are exposed in the eastern half of the quadrangle; their general trend is north-northeast. The metavolcanic rocks in the southwestern quarter of the quadrangle are separated from those in the eastern half by the great peridotite sheet of Josephine Mountain; west of the peridotite the rocks show a less uniform trend but generally strike northwest.

Recent mapping by Wells and Walker in the Galice quadrangle has shown that a thick assemblage of volcanic rocks occurs between the Dothan and Galice sedimentary rocks. They believe that these volcanic rocks were deposited conformably on the Dothan and that in all probability the Galice was deposited on the volcanic rocks conformably. They have named these volcanic rocks the Rogue formation (Wells and Walker, in press). The volcanic rocks in the Patrick Creek-Shelley Creek area and the Craigs Creek-Coon Creek area are probably the stratigraphic equivalent of the Rogue formation.

Although areas of metavolcanic rocks are thoroughly dissected, the surfaces of the ridges are generally smooth, and bold ledges and cliffs are uncommon. Exposures of metavolcanic rocks are rare in the beds of creeks draining areas underlain by these rocks, as the channels are choked with accumulations of blocks, boulders, and gravel.

The thickness of the metavolcanic assemblage of the Galice formation within the Gasquet quadrangle cannot be accurately determined. In the Kerby quadrangle to the north, where the assemblage is better exposed, Wells (1949, p. 7) estimates the thickness to be about 10,000 feet. In the canyon of Rogue River 45 miles to the north, the Rogue formation, which is probably the same sequence of rocks, is 15,000 feet thick (Wells and Walker, in press). If allowance is made for folding and duplication by faulting, some of the sections exposed in the Gasquet quadrangle must be at least 7,000 feet thick.

The lithology of the metavolcanic assemblage of the Galice is diverse. The volcanic nature of the rocks over large areas is easily recognizable, although some of the rocks are schistose, sheared, and chloritized. The rocks comprising the assemblage include meta-andesite, metabasalt, spilite, possibly a little metarhyolite, meta-andesite tuff, metarhyolite tuff, agglomerate and flow breccia, and a few scattered intercalated lenses of black slate. In one place a layer of volcanic agglomerate was mapped as a separate unit; individual flows within a series of flows and tuffs can rarely be distinguished because of metamorphism.

Meta-andesite flows and flow breccias are probably more abundant than any of the other rocks. They are grayish green to dark green; some are porphyritic and some felsitic. Amygdaloidal facies are rather common, but as a rule amygdules are recognizable only by close examination. In places flow breccias are conspicuous. The structure of these flow breccias is best seen on polished, water-worn exposures in creek beds. The rock consists of a heterogeneous mass of blocks, angular pebbles, and scoriaceous fragments in a fine-grained matrix of meta-andesite. In places fragments make up more than half of the flow, but in other places they are more widely scattered. Many of the rocks have a well-formed andesitic (pilotaxitic) texture, but in others the texture is felsitic.

The commonest type of meta-andesite consists of oligoclase or andesine, hornblende or augite as primary minerals, and albite, actinolite, epidote, clinozoisite, zoisite, biotite, sericite, chlorite, quartz, calcite, prehnite, and pyrite as secondary minerals; sphene and magnetite are accessory minerals. Some specimens retain unaltered feldspar which may be either oligoclase or andesine, but in most the plagioclase is saussuritized in varying degrees; in still others the original feldspar has been replaced by albite. Replacement by albite, where present, is generally complete, but in a few specimens part of the more calcic feldspars remain. Hornblende is the most abundant of the mafic minerals; augite is present in subordinate amounts in some specimens but was not found to occur together with hornblende in any of the thin sections examined. As a rule the hornblende and augite have been less subject to alteration than feldspar, but their alteration to chlorite and epidote is common and to biotite less common. As much as 5 percent quartz is present as an original constituent in a number of thin sections, but in most the quartz appears to be a later addition. Some of the rocks contain small amounts of disseminated pyrite. Amygdules consist of calcite, quartz, chlorite, and prehnite.

Metabasalt is much less abundant than meta-andesite. Generally speaking, the colors are darker than those of meta-andesite, but a few are medium gray. Most of these metabasalts have a diabasic texture; they are generally nonporphyritic, but phenocrysts of feldspar as much as 2 millimeters long can be seen in some. Amygdaloidal textures are also common. Metabasalt forms pillow structures in a few places, notably along Craigs Creek and its tributaries and in excellent exposures along the West Fork of Shelley Creek; at the latter locality well-formed pillows 1 to 3 feet in largest dimension are conspicuous features of the flows. The very fine grained margins of individual pillows are highly epidotized.

The primary minerals in the metabasalts recognized under the microscope are labradorite, augite, and hornblende. Little labradorite remains in any of these rocks, and in most of them it has been altered very largely to epidote and zoisite or has been replaced by cloudy albite or sodic oligoclase. Augite is abundant and is in part altered to chlorite; one or two specimens contained aggregates of chlorite that may have been pseudomorphic after olivine. The secondary minerals are oligoclase or albite, calcite, prehnite, actinolite, chlorite, magnetite, epidote, clinozoisite, zoisite, and in some a little quartz. The original mafic minerals in highly altered specimens have been converted to hornblende and epidote; such secondary hornblende is commonly pleochroic in bluish tints.

A few specimens, including some of the pillow lavas, bearing a superficial resemblance to metabasalt are best described as spilite. Spilites are soda-rich rocks of basaltic type in which albite or sodic oligoclase is the predominant feldspar. The term "spilite" was used first by Dewey and Flett (1911) in describing a suite of rocks having these characteristics. Spilites and the spilite problem of eastern Oregon have been described by Gilluly (1935); spilites of the Olympic Peninsula were studied by Park (1946). The spilitic rocks of the Gasquet quadrangle are gray to dark green and rather fine grained; some are amygdaloidal. Megascopic phenocrysts of plagioclase and augite are visible in some specimens. These spilites have diabasic texture, a texture characteristic of fresh rocks whose feldspars are of labradorite or more calcic composition.

Under the microscope the spilite is seen to consist of albite, augite, chlorite, calcite, sericite, prehnite, a little quartz in a few specimens, and rare scattered needles of tremolite or actinolite; epidote and zoisite are notably absent. The albite is cloudy and contains numerous chlorite inclusions. The lack of epidote and zoisite indicates albite was not formed by saussuritization. On these grounds and the demonstrable albitization of some of the meta-andesites, the feldspars in the spilites are believed to be the result of albitization. Augite in some of the specimens is extensively altered to chlorite. Calcite is relatively abundant, especially in amygdules and veinlets.

The spilite and albitized meta-andesite of the Gasquet quadrangle probably evolved under similar conditions. Both rock types are believed to have formed by the reaction of sea water on thick rapid submarine accumulations of lava. Accordingly, basaltic lava formed spilite, and andesitic lava formed albitized andesite. As Park (1946, pp. 318–320) has pointed out, only during periods of rapid extrusion of thick flows is sufficient heat retained by the reacting system to promote the stewing action of trapped aqueous solutions containing salt and the gas diffusion necessary for any considerable alteration. Thin flows would chill too rapidly to permit more than negligible alteration.

Most of the tuffs in the metavolcanic unit are gray or greenish-gray structureless, massive rather fine grained rocks almost indistinguishable in hand specimens from some of the fine grained flow rocks. Most of the tuffs are andesitic, but a few thin layers of rhyolitic tuff were seen. Distinct bedding is rare. Less common are rocks of coarser grain whose pyroclastic nature is evident even when viewed from a distance of several feet. In few of the tuffs, however, are individual fragments larger than a small fraction of an inch across.

Microscopic examination shows that the tuff consists of plagioclase, chlorite, epidote, and clinozoisite, with smaller amounts of quartz, amphibole, augite, calcite, sericite, and magnetite; small lithic fragments are common. The rocks have been extensively recrystallized, forming minerals characteristic of low-grade metamorphism; thus the more calcic feldspars are commonly highly saussuritized, and the mafic minerals are altered largely to chlorite. In some the original fragmental nature of the rocks is almost obscured by alteration.

A distinctive layer of volcanic agglomerate is exposed in the southeast corner of the quadrangle. This layer forms one of the few beds traceable for any distance in the entire quadrangle; it ranges in thickness from about 150 feet to at least 300 feet. The lithology and texture of this layer differ considerably from place to place. Just east of the Gasquet quadrangle, where this layer is exposed in the cuts of the Bear Basin road, the fragments are very coarse—as much as 4 inches across—and angular, and sorting is nonexistent. Elsewhere fragments are smaller, and near Hurdygurdy Butte parts of the bed show rude sorting; some of the fragments are water-worn and rounded. The rock is composed very largely of andesite fragments in a sandy matrix of volcanic debris, but pebbles of quartz and other rocks are common locally, especially in the obviously water-laid parts.

Immediately west of the layer of volcanic agglomerate black slates and sandstones crop out. North of Hurdygurdy Butte these slates and sandstones interfinger with flows and tuffs of the volcanic series. Because of extremely dense brush and very poor exposures, mapping of these interfingers was impossible, and their representation on the map is largely diagrammatic.   (See pl. 12.)

Overlying the metavolcanic rocks is an assemblage of metasedimentary rocks composed principally of slate and phyllite with interbedded tuffaceous sandstone.   One small lens of limestone enclosed in the slates is known to exist in the Gasquet quadrangle.   Sedimentary rocks are exposed in the southwestern and eastern parts of the Gasquet quadrangle.   In the eastern half, one belt trending north-northeastward extends the length of the quadrangle, and two others occupy smaller areas on Monkey Creek and near Hurdygurdy Butte.

Areas underlain by slate and sandstone are characterized by steep smooth slopes devoid of ledges and cliffs.   In general these rocks crop out only in stream channels, and here exposures may be nearly continuous except on those parts of stream channels that are wide, flat, and of low gradient.

The slates and sandstones in the eastern half of the quadrangle apparently are isoclinally folded and in general dip from 45° to 75° E. Along Coon Creek and in Lower Coon Mountain in the southwestern part of the quadrangle slates are folded into a relatively gentle syncline plunging southward.   Elsewhere the scarcity of outcrops and the lack of distinctive beds prevent accurate determination of the major folds.   Drag folds and minor crumples are numerous.   Inasmuch as the number of times the beds are repeated in the isoclinal folds is uncertain, the thickness is unknown; it must be at least 3,000 feet and may be much more.

The slate and phyllite units of the formation are dark gray or black but weather grayish white, yellowish, or brown.   Most of the rocks are well-bedded, and in many places rhythmically so, the strata consisting of thin alternating layers and lenses of black slate and grayish fine-grained sandstone.   In general bedding and cleavage are parallel, and in areas of homogeneous rocks cleavage obscures the bedding. Over large areas the rocks are slaty, but variations in degree of metamorphism are common; some rocks are but slightly metamorphosed, others are phyllitic, and some, in zones of particularly intense deformation, are schistose.   A few of the black highly argillaceous slates possess a peculiar splintery fracture, so that the rock breaks and weathers out in long, thin, pencillike fragments a quarter to one half an inch in diameter and as much as 6 inches long.

Microscopic examination reveals that the slate consists of plagioclase, quartz, chlorite, micas, carbonaceous and argillaceous material, calcite, magnetite, a little hornblende, and scattered rock fragments. Locally the material appears to be graphitic.   The average grain size of the slate is very fine, probably less than 0.005 millimeter, and in some specimens no grains are larger than 0.01 millimeter. In the slates most of the larger grains are angular, but in general, mineral grains are flattened, and recrystallization has resulted in the growth of micaceous minerals; the phyllites are, of course, more highly recrystallized, mineral grains are more highly flattened, and flakes of mica are larger. In one specimen of black slate, quartz grains have grown across thin layers of carbonaceous or graphitic material, and these layers pass through the quartz grains undisturbed. Under the microscope the slates that break into pencillike fragments show closely spaced joints cutting the cleavage planes, and the cleavage planes show drag along joint surfaces. Some slates are fractured, and individual fracture fragments are rotated slightly. Thin sandier lamellae in slate are lighter colored and contain much less black argillaceous material.

The sandstones occur both as thin layers interbedded in slates and as thick, massive beds devoid of slate partings. One massive sandstone bed several hundred feet thick, which crops out in the vicinity of Kelly Peak and the mouth of Jones Creek, was mapped separately. This sandstone layer grades gradually downward into massive, structureless tuffs belonging to the assemblage of metavolcanic rocks, upward into well-bedded sandstone, and from well-bedded sandstone upward into slate. In this locality the sandstone forms the bottom layer of the sedimentary sequence, but it was not seen elsewhere at contacts between the metavolcanic and sedimentary rocks of the Galice formation.

The sandstones are largely fine- to medium-grained rocks of moderately light gray or greenish-gray colors; a few layers of grit occur. In some places the rocks have a distinct fissility analogous to the cleavage in the slates; fragments are flattened, and parting surfaces have a silvery sheen resulting from the formation of sericite. Near the mouth of Dead Horse Gulch and on Griffen Creek near the junction with Smith River are sandstones containing angular fragments of black shale as much as 3 or 4 inches long; the origin of these fragments is unknown.

Under the microscope the sandstones are seen to be highly arkasic and tuffaceous and consist of plagioclase, quartz, augite, hornblende, chlorite, epidote, micas, fragments of volcanic rocks, quartzite, and shale, shards of devitrified glass, and minor amounts of carbonaceous and argillaceous material. Mineral grains and rock fragments are notably angular, and much of the feldspar is fresh. In some specimens a slight schistosity is evident in combination with considerable recrystallization.

One small lens of gray marbleized limestone crops out in the Gasquet quadrangle about three-quarters of a mile west-northwest of the

Higgins triangulation station. This lens is 500 to 600 feet long and 60 to 70 feet thick; it is completely enclosed in slate and fine-grained sandstone. The rock is rather coarsely crystalline, has a slight fissility, and contains considerable sandy and tuffaceous material.

Several small collections of fossils have been made from the Galice formation on Galice Creek, on the Rogue River, and on Graves Creek farther north in Oregon. Diller reported the discovery of a few invertebrate fossils in the Gasquet quadrangle on Shelley Creek near Monumental, and both Taliaferro and Maxson collected fossil plants on Shelley Creek. These fossils indicate the formation is of Late Jurassic age, probably Kimmeridigian, and about the same age as the Mariposa slate of the Sierra Nevada.

CONDITIONS OF DEPOSITION

The conditions under which the sediments of the Galice formation were deposited are not altogether clear. Probably the most notable feature of the sediments is the large proportion of pyroclastic material. Only the most fine grained of the black slates are free of definitely pyroclastic material, and even in these the high percentage of feldspar and chlorite is suggestive of volcanic contributions. Not only is most of the material of volcanic origin, but also it shows no sign of weathering and is practically free of rounded, water-transported material. Therefore, it seems evident that the bulk of the sediment was derived from active subsurface or near-surface volcanoes, and only a small part came from the subaerial erosion of volcanic islands or of the adjacent mainland.

The few fossils collected from the Galice formation give no definite evidence as to whether the sediments were deposited under neritic or bathyal conditions. The great thickness of very fine grained sediments suggests deposition of deep water, and the lack of limestone except for two known reeflike lenses (one outside the Gasquet quadrangle) may indicate that deposition of much of the material was at depths in which calcium carbonate is soluble. Thus, most of the Galice formation is presumed to have accumulated in deep water in a region of active volcanism.

INTRUSIVE ROCKS

Plutonic rocks underlie about six-tenths of the surface of the Gasquet quadrangle. Although these rocks are most widely distributed throughout the western half of the quadrangle and underlie all but about one square mile of the northwest quarter, they also underlie considerable areas in the eastern part of the quadrangle. Most of the masses have pronounced northerly trends in part conformable to the invaded rocks, but over large areas they cut across these invaded rocks at small angles.

The plutonic rocks have been divided on a basis of composition into two groups: the ultramafic rocks and the gabbro and hornblende diorite rocks. The older, and by far the more widespread, are the ultramafic rocks. From a point 2½ miles south of the Gasquet quadrangle the great peridotite sheet of Josephine Mountain trends northward across the length of the quadrangle and into Oregon to a point at least 40 miles north of the California boundary. This peridotite sheet is the largest body of ultramafic rock in the United States and perhaps in North America. Other ultramafic bodies occupy large areas on Lower Coon Mountain and along the South Fork of Smith River in the southwest part of the quadrangle, around Table Mountain and Blue Ridge in the southeast part of the quadrangle, and north of Monumental in the northeast corner of the quadrangle. Gabbro and hornblende diorite is widely distributed but much less extensive than the ultramafic rocks; the largest bodies are in the northeast part of the quadrangle and between Gordon and Hurdygurdy Creeks in the south-central part of the quadrangle. Numerous smaller bodies occur elsewhere.

ULTRAMAFIC ROCKS

GENERAL FEATURES

Ultramafic rocks are igneous rocks composed essentially of one or more of the common magnesium-iron silicate minerals—olivine, pyroxene, and amphibole—and are characterized by a near or complete absence of feldspar. Ultramafic rocks composed of olivine only and of pyroxene only are known as dunite and pyroxenite, respectively. The rock of intermediate composition containing both pyroxene and olivine is called peridotite, of which the varieties saxonite, lherzolite, and wehrlite are present in the Gasquet quadrangle. Small amounts of amphibolite, a rock composed essentially of amphibole, also occur; inasmuch as this rock appears to be the result of alteration rather than of primary igneous origin, it will not be discussed in this section.

The ultramafic rocks intrude metavolcanic and sedimentary rocks of both the Galice and Dothan formations largely as rudely conformable masses. Most of the varieties are serpentinized to some degree and in places, especially along shear zones and contacts, are completely serpentinized. In these areas the original composition of the rocks cannot be determined. So widespread is serpentinization that all the ultramafic rocks are called serpentine by the local inhabitants.

Areas underlain by ultramafic rocks can be recognized from a great distance because of the scarcity of vegetation and the red color of the outcrops. Vegetation is scanty because the soils developed from these rocks are commonly poor, thin, and droughty. The droughty condition is due to the highly fractured nature of the ultramafic rocks which permits deep rapid drainage of the soil and near-surface rocks.

Zones of highly sheared serpentine, known as "slickentite," are distinguishable by the almost complete lack of vegetation and the light grayish-green color of the rock.

The topography developed on the ultramafic rocks is rough, except for the broad, gently rolling plateau surfaces that are remnants of the Klamath peneplane. Even on these remnants the soil covering is scant, and the surface is covered with loose boulders and jagged ledges. One of these surfaces is shown in the foreground of plate 12. This blocky, uneven surface is the result of well-developed joint systems and the irregular veins and seams of serpentine that favor differential erosion. Streams draining peridotite areas maintain fairly uniform flows the year round because of the high permeability of the rock and the low surface runoff.

SAXONITE

Saxonite is by far the most abundant of the ultramafic rocks and underlies about one-half of the surface of the quadrangle. Except for the two large sills of wehrlite, the bulk of the peridotite masses, including the peridotite sheet of Josephine Mountain, is composed of this rock.

Saxonite is a rather coarse grained rock composed of olivine and the orthorhombic pyroxene enstatite. The grains of enstatite are 2 to 5 millimeters in diameter and the grains of olivine are, as a rule, slightly smaller. In a few areas the saxonite has a distinct banded or layered appearance due to alternating enstatite-rich and enstatite-poor layers. These layers are commonly about half an inch thick. Olivine and enstatite both alter to serpentine; enstatite commonly alters to bastite, a serpentine material preserving the original form and cleavage of enstatite. Both enstatite and bastite weather less readily than either fresh or serpentinized olivine; consequently, weathered surfaces of saxonite are rough and studded with projecting crystals of enstatite or bastite. (See pl. 13, A.) Weathering of saxonite liberates iron oxide which stains the rock surfaces a deep reddish brown. Fresh surfaces of saxonite are very dark green or nearly black if serpentinized and very pale green and translucent where not serpentinized. Outcrop areas of saxonite are generally rough; the rock breaks down along joints or along seams of serpentine.

Under the microscope the saxonite is seen to consist of olivine and enstatite and a little accessory magnetite and chromite as primary constituents. Secondary antigorite, bastite, and magnetite dust are always present. Common, but not present in all specimens, are chrysotile, talc, chlorite, and rarely a little tremolite. Enstatite apparently alters more readily to serpentine than olivine does; in sections in which olivine is less than half altered the enstatite may be entirely serpentinized.

Dunite occurs as a minor, but widely distributed, facies of the ultramafic masses, especially those composed largely of saxonite. No dunite was found in the wehrlite body capping Lower Coon Mountain, and only very small amounts were seen in the wehrlite east of Hurdygurdy Butte. Bodies of dunite range in size from highly irregular clots a few inches long to lenticular masses several hundred feet long. Masses of considerable size are commonly elongate in the direction of the major axis of the ultramafic intrusive. Contacts between dunite and the enclosing saxonite are, as a rule, rather sharply defined, but in places the saxonite grades into dunite by decreases in the amount of enstatite. No attempt was made to delineate dunite bodies in field mapping because of the time involved and the small size of individual bodies. The chromite deposits are found only in dunite.

The dunite is a medium- to coarse-grained rock that differs conspicuously in outcrops from saxonite. Weathered surfaces of dunite are smooth and have a yellowish or buff color which contrasts with the rough surfaces and reddish colors of saxonite. (See pl. 13, A.) Like saxonite, fresh surfaces of dunite are very dark green or nearly black where serpentinized and light green and translucent elsewhere. Serpentinized dunite has a smooth conchoidal fracture.

The microscope shows dunite to be composed of a xenomorphic mass of olivine grains and small amounts of accessory chromite. Many of the chromite crystals are well-formed octahedra. All the dunite has altered in some degree to antigorite, but as a rule the amount of secondary magnetite formed is very small. This probably indicates the olivine is low in iron. A small amount of talc is present in some thin sections.

LHERZOLITE

Lherzolite, a rock composed of olivine and both orthorhombic and monoclinic pyroxene, occurs in unknown, but probably small, amounts in the saxonite. In hand specimens and in outcrop the appearance of the two is identical; only under the microscope can they be differentiated. Under the microscope olivine and various amounts of both enstatite and diallage can be identified. Both varieties of pyroxene alter to bastite apparently with greater ease than olivine alters to antigorite. Considerable amounts of secondary magnetite are liberated in the serpentinization of diallage, but the amount of secondary magnetite liberated by the serpentinization of enstatite is small, and the amount of magnetite formed by the serpentinization of olivine is even smaller.

WEHRLITE

Two sill-like masses of wehrlite, a rock composed of diallage and olivine, crop out within the quadrangle. One mass intrudes slates of the Galice formation and caps the western end of Lower Coon Mountain; the other intrudes metavolcanic rocks of the Galice formation at the edge of the quadrangle east of Hurdygurdy Butte. The mass on Lower Coon Mountain is more conspicuously sill-like than any other large plutonic body in the quadrangle; it occupies the trough of a southward-plunging syncline. The form of the mass east of Hurdygurdy Butte is less well known, as the region to the east is unmapped and only the western margin laps over into the quadrangle, but this margin is distinctly conformable.

Wehrlite is a dark very coarse grained rock having crystals half an inch or more in diameter. Weathered outcrops have an extremely rough surface and a reddish color. Fractured surfaces are very uneven and hackly. On the average pyroxene makes up about 75 percent of the rock—thus it grades into pyroxenite—but grades from as little as 50 percent pyroxene into pyroxenite that contains 100 percent. Soils developed from wehrlite are thicker, apparently more fertile, and commonly support much denser vegetation than do soils derived from other varieties of ultramafic rocks.

Under the microscope wehrlite is seen to consist of diallage and olivine, accessory magnetite, and, in one thin section, a little sphene as primary minerals; antigorite, chrysotile, actinolite, chlorite, and magnetite are common secondary minerals; a small crystal of biotite was present in one section. The texture of the rock is xenomorphic. Diallage has lobed edges against olivine and appears to have partly replaced it. Some of the diallage has a moth-eaten appearance due to the formation of secondary actinolite in optical continuity with the diallage, but there is only slight alteration of diallage to serpentine minerals. Olivine in wehrlite alters to antigorite much more readily than does diallage, a reversal in ease of alteration of the two minerals in the lherzolite. Expansion fissures are rather common in diallage adjacent to serpentinized olivine grains. At two places, one on the trail just above Coon Creek near the eastern edge of the Lower Coon Mountain mass, and the other at the southern edge of the quadrangle near the western contact of the eastern mass of wehrlite, specimens were collected that contained as much as 30 percent interstitial plagioclase. The plagioclase in one specimen was An_{62} in composition and in the other was as calcic as An_{72}. The plagioclase-bearing rocks have gabbroic compositions, but their affinities are clearly those of the wehrlite rather than of the later diorite-gabbro intrusions.

PYROXENITE

Pyroxenite is the least abundant of the ultramafic rocks. Bodies of pyroxenite are small and of only local occurrence. Scattered lenses of pyroxenite, a few inches to a few feet across, composed of enstatite occur only as local facies in saxonite. These lenses are common along the old Wimer road in the northeast corner of the quadrangle. Small masses of diallage pyroxenite are rather common in wehrlite.

The pyroxenite is a very coarse grained rock with crystals as much as an inch long. The color on fresh surfaces is very dark green or nearly black and the fracture is extremely rough and hackly. Of all the ultramafic rocks pyroxenite is the most resistant to weathering, and outcrops ordinarily project above the ground surface as rough, jagged ledges. Pyroxenite is much less susceptible to serpentinization than other ultramafic rocks.

SERPENTINE

All the ultramafic rocks, except possibly some small areas of pyroxenite, are serpentinized to some extent. Serpentine is most abundant around the margins of the saxonite masses and along shear zones in these masses. Most of the small ultramafic bodies everywhere and all the very small ultramafic bodies intruded along faults are serpentine. Only those areas where the rock is nearly or completely serpentinized, where evidence of the original composition of the rock is largely destroyed, were mapped as serpentine. Contacts between highly serpentinized areas and relatively fresh ultramafic rocks are gradational, and therefore on the map the contacts are indicated by unmarked boundaries.

Most serpentine is grayish green but varies from light green to black; it is relatively soft and very incompetent. Highly sheared serpentine, or slickentite, is composed of a mass of curved flattened semitranslucent greenish to honey-colored flakes; the surfaces of these flakes are invariably slickensided. In general, rounded pebbles or boulders of unsheared serpentine are scattered through the mass of highly sheared material. In areas of intense alteration considerable talc forms from serpentine. Opal and magnesite are common products of weathering.

EMPLACEMENT OF THE ULTRAMAFIC ROCKS

In peridotite, as a typical example of most ultramafic rocks, the temperatures of crystallization are high for all minerals except the hydrous magnesium silicates of the serpentine group; contact metamorphic and thermal effects caused by the intrusion of these ultramafic rocks, however, are notably weak and in most places nonexistent. Many explanations of these abnormal facts have been proposed. For example, it has been argued (Bowen and Schairer, 1936, pp. 395-396) that large bodies of mush consisting of crystals of ultramafic minerals and a small percentage of the parent liquid were carried along by the complex parent magma and intruded where now found; and that during final emplacement the interstitial liquid which had cooled far below the temperature of crystallization of the ultramafic minerals was squeezed out by filter pressing to yield rocks of ultramafic composition. This parent magma, however, has left no corpus delicti in this area, and there is a complete absence of any trace of the postulated interstitial parent liquid within the existing peridotites. Such perfection of filter pressing strains the credulity of the most willing believer. Furthermore, there is abundant evidence that some of the olivine crystallized after the peridotite bodies came to rest. The presence of an olivine liquid phase indicates that the temperature of the crystal mush could not have been much less than the temperature of crystallization of olivine. The foregoing evidence seems to preclude the existence of a liquid fraction of low crystallization temperature either within or associated with the intruding peridotite bodies, which in the case of the Josephine Mountain mass consisted of over 500 cubic miles of peridotite. Therefore it must be assumed that the magma was intruded at high temperature and another explanation of the absence of contact metamorphic and thermal effects should be sought.

The only phenomenon almost universally associated with the peridotite intrusions is extreme shearing along the contacts; this shearing involves both the intruded rock and the peridotite, but especially the peridotite. It is probable that although the contacts are intensely sheared, no significant displacement has occurred along many of them since final emplacement of peridotite. The lack of contact metamorphic and thermal effects and the pervasive shearing of contacts suggests a peculiar mode of intrusion. The writers conclude that the initial mass of essentially anhydrous magma absorbed water in its journey from its source. That water will move from a saturated porous rock into an anhydrous magma has been demonstrated by Sosman (1950). He writes:

Laboratory data and plant experiments with blast-furnace slag have demonstrated the solubility of water in molten silicates, even under atmospheric pressure. An intruding magma, if unsaturated with respect to water, will therefore establish a gradient of both water pressure and water concentration in surrounding rocks; the direction of falling pressure and concentration will be toward the intrusive, not away from it as has been commonly assumed. This gradient is further enhanced by the phenomenon of thermal transpiration, by which a gas under constant pressure, in a medium having small pores, travels toward the region of highest temperature.

The highly incompetent serpentine shell was crushed and sheared as it was carried along by the advancing magma. Undoubtedly, the hot interior magma would break through the relatively cool serpentine shell at times, but such escaping magma probably would soon be reduced to serpentine and incorporated in the shell. When the magma came to rest it was enclosed within an envelope of slickentite, which insulated the wall rocks from the effects of the hot, partly liquid, interior.

GABBRO AND HORNBLENDE DIORITE ROCKS

Several fair-sized bodies of gabbro and hornblende diorite crop out in the quadrangle. The largest of these is exposed between Monkey Creek and Monkey Creek Ridge, and in the headwaters of Monkey Creek. Two other closely related masses crop out near the head of Diamond Creek where a small mass of quartz diorite also crops out. All these masses may possibly be connected at depth. A long, relatively narrow, dikelike mass intrudes the ultramafic rocks north of High Plateau Mountain. In the southwestern quadrant of the quadrangle other masses crop out at the mouth of Coon Creek, below French Hill on Craigs Creek, and between Gordon and Hurdygurdy Creeks. Smaller masses, most of them too small to be shown on the map, are numerous elsewhere, especially in the metavolcanic rocks and along the contacts of ultramafic bodies. Almost invariably where gabbro or hornblende diorite intruded ultramafic rocks the intrusions took place along highly serpentinized zones, probably because these were zones of weakness.

The gabbro and hornblende diorite appear to be facies of a single magma, and they intergrade. Because of poor exposures and the complex facies relations between the two types, no satisfactory distinction could be made between them within a given intrusion. Consequently, intrusives were mapped as hornblende diorite where that rock was the dominant type and gabbro where gabbro was dominant; in the larger masses both rocks invariably appeared to be present. In many hand specimens the two rock types are indistinguishable. In general hornblende is the dominant mafic mineral in the diorite, and augite in the gabbro; however, much of the hornblende diorite also contains appreciable quantities of augite. Most of the rocks are somewhat altered, and in some the feldspar is completely saussuritized. The dark minerals were less affected by alteration, but chlorite and epidote formed from them are common.

Unlike the ultramafic rocks, the gabbro and hornblende diorite magmas were "wet" and produced decided changes in the enclosing rocks. The brittle metavolcanic rocks were most susceptible to alteration, and in many places the contact zones between these rocks and gabbro and hornblende diorite are broad and gradational, especially below the French Hill mine on Craigs Creek and on Shelley Creek about half a mile above its mouth. Schlieren of country rock in various stages of digestion are common near the borders of the larger masses. In a few local areas near contacts, principally in the gabbro and hornblende diorite mass in and near the Monkey Creek drainage basin, the rock has a vague gneissic structure, but nowhere is it sufficiently well marked to warrant classing the rock as a gneiss.

GABBRO

The gabbros are chiefly gray medium-grained crystalline rocks; plagioclase, and augite or amphibole, in grains 1 to 3 millimeters in diameter are the only minerals identifiable in hand specimens. Very small light-colored veinlets of zoisite and prehnite are common. Most of the gabbro has a hypidiomorphic texture, but in some specimens cataclastic modifications are common, and in others the texture is crystalloblastic.

Under the microscope plagioclase, hornblende, augite, and accessory magnetite and sphene can be seen as primary minerals, and actinolite, tremolite, uralitic hornblende, chlorite, zoisite, epidote, sericite, albite, calcite, pyrite, and prehnite as secondary minerals.

Most of the plagioclase is labradorite, but in some specimens plagioclase as calcic as anorthite is found. The composition of the plagioclase ranges from An_{50} to An_{83}. Zoning is commonly absent. All the plagioclase has been altered at least in part to saussurite, and some of it has been completely converted. Augite or its alteration products commonly make up about one half of the rock. In some specimens augite has lobed edges against plagioclase and appears to have been extensively replaced by that mineral. Reaction rims of hornblende, uralitic actinolite, and rarely a little tremolite, invariably replace some of the augite. Hornblende is present in some, but not all, specimens, and where present it may constitute as much as 10 percent of the rock. In some specimens ripidolite, a variety of chlorite, is a very common mineral. Epidote and zoisite are present in most specimens. Prehnite occurs in a few of the more highly altered rocks.

HORNBLENDE DIORITE

Like gabbro, hornblende diorite is commonly a gray medium-grained crystalline rock but is more variable in both composition and texture than gabbro. The proportions of light and dark minerals vary widely; in some specimens the amount of light-colored minerals may be as low as 20 percent and in others as high as 70 percent. In some specimens all mineral grains may be less than 1 millimeter across, and in others a few crystals may be almost 1 centimeter across.

Under the microscope the diorites are seen to consist of plagioclase, hornblende, and augite as primary minerals; sphene, magnetite, more rarely apatite, and perhaps ilmenite, are accessory minerals. The identification of ilmenite is doubtful. Secondary minerals are uralitic hornblende, epidote, clinozoisite, zoisite, chlorite, prehnite, albite, calcite, leucoxene, biotite, sericite, and pyrite.

The plagioclase has a compositional range from An_{20} to nearly An_{50}. Much of the plagioclase is oligoclase, and the corresponding rocks are sodic diorite. In some specimens two kinds of plagioclase are present, andesine and albite, and the rock appears to have acquired a considerable secondary addition of soda. Zoning is not common but is present in some crystals. Alteration to saussurite is pervasive, and in some rocks no original feldspar remains. Hornblende is the most abundant primary mafic mineral, but augite is present in some specimens. Where augite is present, it invariably has been replaced to some extent by either hornblende or actinolite. Part of the augite has been altered to a rusty mass of indeterminate composition. Rarely, a little biotite forms along the cleavage planes of hornblende. Chlorite, epidote, and zoisite are common, and some rocks are veined with prehnite. Scattered grains of pyrite are rather common.

Most of the diorite has a hypidiomorphic granular texture, but this texture shows cataclastic modifications in some specimens, and in still others the texture is crystalloblastic. In general the secondary minerals of both the diorite and the gabbro are those characteristic of low-grade metamorphism, although the rocks have by no means reached equilibrium.

QUARTZ DIORITE

The only quartz diorite found in the Gasquet quadrangle crops out on the North Fork of Diamond Creek near the north edge of the quadrangle and as small dikes at the Webb Cinnabar mine. Along the North Fork of Diamond Creek the quartz diorite occurs as irregular masses intruded into a complex contact zone between hornblende diorite and serpentine; in fact, so large a part of this contact zone consists of quartz diorite that it was mapped as such. In addition to quartz diorite the contact zone contains numerous dikes of other rocks, most of which appear to be related to the hornblende diorite, although some may have been formed considerably later and genetically unrelated to the hornblende diorite. The quartz diorite is thought to be a highly silicic differentiate of the main hornblende diorite body intruded near the end stages of its emplacement.

The quartz diorite is a light-colored medium-grained crystalline rock. The texture of the rock is hypidiomorphic and shows some crystalloblastic modifications. Minerals recognized in hand specimens are plagioclase, bluish quartz, and chlorite. Chlorite forms folia, several millimeters across, which tend to wrap around quartz and feldspar grains.

Microscopic examination shows the rock to consist of andesine, quartz, hornblende, and accessory apatite and magnetite as primary minerals; chlorite, epidote, and sericite are secondary minerals. The andesine is about half altered to saussurite, and the quartz shows wavy extinction and is partly recrystallized. Chlorite appears to have been derived largely from biotite, but some has been derived from the alteration of hornblende.

AMPHIBOLITE

The amphibolite of the Gasquet quadrangle has resulted from the alteration of serpentine by intrusions of gabbro and hornblende diorite. Amphibolite occurs as clots and xenoliths in gabbro and hornblende diorite near serpentine contacts and is found at scattered localities wherever gabbro and hornblende diorite intrude serpentine. Individual masses are rarely more than a few feet long and are commonly only a few inches long. Serpentine inclusions are not always completely altered, and, indeed, some are outwardly little affected by engulfment in the later magmas.

Amphibolite is a dark-green coarse-grained rock in which only amphibole is recognizable in hand specimens, but some specimens show films of a light-colored material coating the amphibole grains. The amphibole of which the rock is composed may be either hornblende or actinolite and tremolite. The actinolite and tremolite are intergrown, the two minerals being in optical continuity but having slightly different extinction angles. One specimen composed of actinolite and tremolite, collected near the Cleopatra mine, also contained minor quantities of prehnite, chlorite, and a little magnetite. Some of the amphibolite is partly altered to chlorite.

DIKES

Dikes consisting of diorite and quartz diorite porphyry, aplite, and rodingite may possibly be related to the gabbro and hornblende diorite, and some undoubtedly are so related; one granite dike, however, apparently is not related to other igneous rocks in the quadrangle. The texture or composition, or both, of all these dikes are sufficiently different from the gabbro and hornblende diorite to warrant their separate description. In general the dikes are confined to the metavolcanic rocks, but some are found in the slate and along contacts of ultramafic rocks. Most of the dikes are clearly pre-Tertiary, but a few rhyolite porphyry dikes are probably of Tertiary age.

The inadequacy of exposures, the scale of mapping, and the time available did not permit showing most of the dikes on the geologic map. Generally the dikes are less than 20 feet thick, and not many can be traced more than a few hundred feet along the strike. Some of the dikes parallel the regional trend of the enclosing rocks; others are intruded along cross faults.

The diorite and quartz diorite porphyries differ essentially only in the absence or presence of a little quartz in the groundmass. The plagioclase is commonly either albite or oligoclase but may be as calcic as andesine; plagioclase phenocrysts are common. Hornblende is the most abundant mafic mineral and in porphyritic rocks is common not only in the groundmass but also as phenocrysts. Augite is sparingly present in some specimens and as a rule shows some alteration to uralitic hornblende. In general the dikes are as highly metamorphosed as the gabbro and hornblende diorite; the feldspars are somewhat saussuritized, and the dark minerals are partly altered to chlorite. Albitization is apparent in some specimens, but in others albite appears to be a primary constituent. Epidote, clinozoisite, zoisite, sericite, calcite, and prehnite are present in some specimens.

The rare aplite dikes have essentially the same composition as the diorite and quartz diorite dikes and differ markedly only in texture.

Probably the most interesting of the dike rocks, because of its peculiar characteristics, is rodingite. It is commonly a fine-grained white or light-green very hard rock that superficially resembles aplite. Rodingite dikes are restricted to the ultramafic rocks. The parent rocks are diorite or gabbro, and rodingite originates from a little-understood interaction between the diorite and gabbro and the peridotite or serpentine wallrock. In many places rodingite can be traced directly from a rock devoid of any of the characteristics of diorite or gabbro into fresh, unaltered diorite or gabbro. It tends to form in the dead ends of dikes or where dikes are thin. Few rodingite dikes are more than 3 feet thick, and most of them form either irregular or lenslike bodies; simple tabular forms are rare.

Rodingite has a rather wide composition range; some types consist largely of grossularite and others of epidote, zoisite, or diopside, but ordinarily all these minerals are present. Common, but less abundant, are pectolite, calcite, and sphene. Rodingites show a manyfold increase in the concentration of calcium over that of the parent gabbroic or dioritic magma, a small relative increase in alumina, a complete loss of potash and soda, and, oddly enough, an almost complete loss of magnesium and iron, the principal cations in the minerals composing peridotite and serpentine. As no determinable exomorphic effects have been found in the wall rocks bordering the rodingites, and in view of the fact that in contact reactions minerals tend to form that equalize the composition of the intruded and intruding rocks, it is, perhaps, possible that this change in the composition of the gabbro or diorite magma to rodingite is due to the loss through the relatively impermeable equilibrium walls of some components of the magma and the retention of others. This is effected by some type of straining, exosmosis, or ultra-filtration rather than by factors that control chemical reactions. This deduction is supported by the fact that rodingites characteristically occur in zones of highly sheared serpentine.

A granite dike in the extreme southeastern corner of the quadrangle, at the contact between wehrlite and metavolcanic rock, may be related to granodiorite that crops out in the vicinity of Preston Peak east of the Gasquet quadrangle. This rock consists of quartz, microcline, orthoclase, oligoclase, biotite, and a little secondary sericite and chlorite; the texture is cataclastic.

TERTIARY ROCKS

SUPRACRUSTAL ROCKS

UPPER MIOCENE BEDS

Remnants, a few acres in extent, of Diller's Wymer beds (1902, pp. 31–35) are exposed at an altitude of 2,000 feet along the Big Flat-Bear Basin road just north of French Flat and at an altitude of 2,200 feet on Lower Coon Mountain. These are the only known exposures of the Wymer beds in the Gasquet quadrangle, but larger areas are exposed about one mile west of Humbolt Flats outside the quadrangle. Other outcrops may exist in the vicinity of French Flat, but search for them in the dense brush and under heavy soil cover was unfruitful.

The beds are marine and consist of friable yellow shale and siltstone that weather red. No fossils were found in the exposures in the Gasquet quadrangle, but imprints of leaves and casts of shells are abundant elsewhere. Fossil collections indicate these beds are of late Miocene age (Diller, 1902, pp. 31–35).

GRAVEL DEPOSITS

A number of deposits of gravel somewhat younger than the upper Miocene sediments are exposed at Humbolt Flats, French Flat, Haines Flat, and on Lower Coon Mountain.

The deposits are poorly sorted and consist of everything from clay to boulders a foot in diameter. Pebbles and boulders are well-rounded and include a large variety of rock types, among which are peridotite, metavolcanic rock, diorite, quartz diorite, gneiss, as well as pebbles of quartz. In places the pebbles and boulders are remarkably fresh and hard, but in others they have decayed thoroughly and crumble at the tap of a hammer. Some of these deposits have an exposed thickness of at least 30 feet, and, according to local placer miners who have worked them for gold, some are at least 70 feet thick. No fossils, other than unidentifiable fragments of wood, were found in any of the gravel. The gravel is deposited at nearly the same altitudes as the upper Miocene sediments. In places the gravel fills channels cut into the upper Miocene beds. In other places, for example at French Flat, these channels cut into the underlying Jurassic rocks, but in no place are they cut very far below the level on which the upper Miocene beds were deposited. Therefore, the gravel is believed to have been deposited shortly after emergence of the upper Miocene beds but before emergence had progressed very far. These relations suggest that the gravels are late Miocene or perhaps early Pliocene.

INTRUSIVE ROCKS

The only intrusive rocks in the Gasquet quadrangle believed to be of Tertiary age are a few scattered albite rhyolite prophyry dikes that cut the peridotite in the vicinities of Diamond Creek and the Cleopatra mine and near the eastern edge of the quadrangle on the Middle Fork of Smith River; some kaolinized dikes, probably of similar original composition, are exposed around the Webb cinnabar mine. These dikes, in general, trend slightly west of north. Most of them are only a few feet thick and can rarely be traced more than a few scores of feet along the strike; the largest one, however, can be traced for about 2 miles and has a maximum width of nearly 400 feet. Some of these dikes have followed the same fractures intruded earlier by Mesozoic dikes of diorite and gabbro.

The rocks forming these dikes are white, light-gray, or pink porphyries, the groundmasses of which are fine-grained and show trachytic textures. The phenocrysts of feldspar are 3 or 4 millimeters across; some phenocrysts are water-clear sanidine.

Under the microscope the rocks are seen to consist of albite, quartz, sanidine, oligoclase, orthoclase, siderite, calcite, garnet, and sericite. Albite in small irregular laths is restricted to the groundmass and comprises from 80 to 85 percent of the rock; quartz, too, is found only in the groundmass and comprises another 5 to 7 percent. Phenocrysts are orthoclase, sanidine, or oligoclase; some of these phenocrysts are single crystals, but others are small aggregates of megascopic crystals. Calcite and rare sericite are secondary, but a little siderite found in one section may have been primary. The siderite has been slightly oxidized and forms rusty spots.

No direct evidence concerning the age of these rocks was found, but the absence of alteration, especially of a mineral as unstable as sanidine, is suggestive of comparatively young rocks. The mineral composition is unlike that of any of the other intrusive rocks in the quadrangle, and the rocks appear to have escaped the deformation to which the other rocks in the quadrangle were subjected.

QUATERNARY ROCKS

The Quaternary rocks, which were mapped as a single unit, consist of unconsolidated terrace deposits, stream sand and gravel, and talus debris. Terrace deposits are relatively abundant along the Smith River but are absent along most of the smaller streams. Some of the terrace deposits are as much as 75 feet above the present stream channels; a few are nearly 75 feet thick. They consist largely of unsorted coarse angular or subrounded debris that contains large blocks as much as 4 or 5 feet in largest dimension. In places gravel consisting of well-rounded pebbles is found (pl. 13, B); these gravels are especially common in the top 2 feet of the deposits along Smith River and on the terraces bordering tributaries such as Monkey Creek. These terraces probably represent relatively short pauses in canyon cutting.

STRUCTURE

The structural features of the rocks of the Gasquet quadrangle are complex and are the results of deformation during at least two periods. The earlier deformation affected only the pre-Tertiary rocks and was probably a result of the Nevadan disturbance at the close of the Jurassic period. This deformation was chiefly compressional and resulted in close overturned folds and high-angle reverse faults. The axial planes of the folds and the associated faults are roughly parallel, trend north-northeast, and dip steeply eastward. Other faults, some trending northeastward and others northwestward, postdate the folds and reverse faults but probably belong in general to the same period of deformation.

The later period of deformation probably began in late Miocene and is still continuing. This later deformation has resulted largely in differential vertical movements and gentle tilting; compressional stresses appear to be absent. No faults clearly referable to the later deformation have been found within the Gasquet quadrangle, but bordering the coastal plain a few miles west of the quadrangle, Tertiary faulting is conspicuous. The drainage pattern of the area is in part controlled by the trend of the older deformation, but the topographic forms clearly reflect the later period of uplift.

STRUCTURAL FEATURES OF PRE-TERTIARY DEFORMATION

The general north-northeast trends, steep dips, and degree of metamorphism indicate deformation by strong compressional forces directed from the east-southeast. Unfortunately, the relations between the Galice and Dothan formations are obscured by intervening intrusions; the interpretation of the structure of the Galice formation is uncertain, especially in the eastern part of the quadrangle. This uncertainty is due partly to scarcity of outcrops, partly to lack of distinctive marker beds, and partly to obscurity of layering or bedding in the thick sequence of volcanic rocks.

Throughout northwestern California and southwestern Oregon the general dip of the strata is southeast. In a normal sequence of beds dipping in this direction the older formations would crop out progressively farther northwest; however, in this region, except for the rocks of the Dothan formation which lie west of a profound structural disturbance, the reverse is true, and the older formations crop out progressively farther to the southeast. Simple overturning would account for reversed sequences in relatively narrow belts but would be insufficient to account for a reversed sequence measured in scores of miles across the strike. An overturned fold capable of producing this phenomenon would require a thickness of strata measured in tens of miles, and the amplitude of the fold itself would have to be even larger. Reversals of sequence result from the thrust faulting that is known to be common in the region, but for this faulting to produce so systematic and widespread a phenomenon appears overly fortuitous. The observed structure and distributions of the formations are most readily explained by the postulation of a series of isoclinal folds on a much larger structure that rises progressively higher eastward and may possibly be the west limb of an anticlinorium. The planing off of this structure exposes older formations eastward, even though easterly dips prevail throughout the region. Thus, although all the rocks have a general steep easterly dip, the rocks of the Gasquet quadrangle are younger than the pre-Jurassic rocks exposed a few miles east of the quadrangle.

FOLDS

COON CREEK SYNCLINE

The most obvious, readily visible fold in the Gasquet quadrangle is a syncline that trends northwestward across the west end of Lower Coon Mountain. The fold is asymmetrical, the east limb being the steeper, and plunges gently southward. The structure of the fold is complicated by the intrusion of a large wehrlite sill and a number of smaller intrusions of gabbro, diorite, and wehrlite. The west limb in the vicinity of the South Fork of Smith River is cut by a large mass of peridotite and a smaller, but fair-sized, body of diorite; both these masses have undergone considerable shearing. Northward, and below the slate the form of the fold is little known because of the lack of visible structure in the metavolcanic rocks. The gabbro in Craigs Creek may be intruded along a northeasterly-trending fault and may indicate the syncline is cut by the prolongation of the fault that trends across the north end of Lower Coon Mountain to the mouth of Patrick Creek and beyond. The anticline that should lie between the Coon Creek syncline and Hurdygurdy syncline, which is the next syncline to the east, is occupied by the great peridotite mass of Josephine Mountain. Along the Smith River near Gasquet this mass plunges southward, roughly conformable to the intruded overlying metavolcanic rocks.

Folds in the eastern part of the quadrangle are complex and not clearly undersoood. However, the belt of Galice sedimentary rocks lying between the peridotite mass of Upper Coon and Gordon Mountains on the west and the peridotite mass of Higgins Butte and Table Mountain on the east, and extending from Smith River southward, is believed to be folded into a tight syncline—the Hurdygurdy syncline—overturned to the west-northwest. The peridotite mass of Upper Coon and Gordon Mountains has destroyed or concealed any structures intervening between the Coon Creek syncline and the Hurdygurdy syncline, at least to the present depths exposed by erosion. North of Smith River in the bed of Shelley Creek the dips decrease westward toward the peridotite and are suggestive of the east limb of an anticline. The belts of sedimentary rocks north of Smith River in Monkey and Griffen Creeks are probably faulted segments of the west limb of the Hurdygurdy syncline. The relationship of the interbedded sedimentary and metavolcanic rocks in the vicinity of Hurdygurdy Butte is unknown, but they may represent lower beds of the overturned east limb of the syncline.

MINOR FOLDS

Minor folds, including drag folds, are numerous in the slates but are rarely found in the more competent metavolcanic rocks. These minor folds range from a few feet to a hundred feet or more across, and commonly the opposite limbs are pressed tightly together. In a tributary of Griffen Creek 1½ miles northeast of Monkey Creek Lookout are a series of small folds plunging vertically. Indicative of stretching that accompanies folding is a small diorite sill about 6 inches thick that crops out near the bottom of Jones Creek about 500 feet above the place where it empties into Smith River. This sill exhibits a well-formed boudinage structure (Cloos, 1947) consisting of an alternate bulging and thinning of the sill into oval-shaped masses. (See fig. 10.) The thin areas are commonly crossed by transverse fractures filled with quartz. Where the sill is completely pulled apart into separate "boudins" the ends of these "boudins" are concave.

The structure of the Dothan formation exposed along the north edge of the northwest part of the Gasquet quadrangle is not determinable within the quadrangle, but immediately northward, in the Kerby quadrangle, an anticline was mapped by Wells (Wells and others, 1949) and called the Baldface anticline. The Dothan rocks in the Gasquet quadrangle occupy the southward extension of the crest of this anticline, which plunges southward under the peridotite. In this area the Dothan formation and the intruded peridotite appear to be mainly conformable, but faults complicate the relationships considerably.

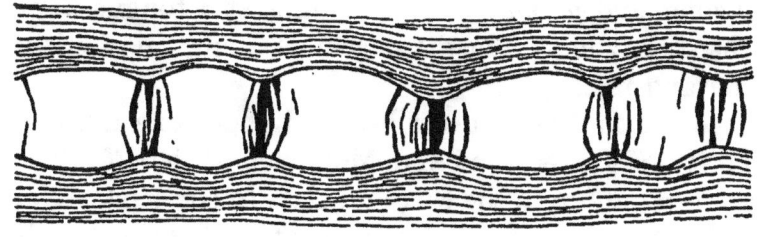

I foot

EXPLANATION

Diorite   Quartz   Metatuff

FIGURE 10.   Sketch of a diorite sill showing boudinage structures.

FAULTS

Faults cutting the rocks of the Gasquet quadrangle are certainly much more numerous than the map indicates. Only those faults displacing well-marked lithologic units and a few visible in actual exposures were mapped; other large faults suspected to exist were not mapped because their existence could not be proved or because their exact location was uncertain. Numerous faults of small displacement were not shown on the map because the scale did not justify plotting them; undoubtedly, multitudes of similar faults were never seen.

Three different systems of the major faults are recognized. The oldest of these systems comprises a set of high-angle reverse faults trending northward and northeastward, dipping eastward, and closely paralleling the regional structure of the rocks involved. The most prominent fault belonging to this system—the southward extension of the Illinois valley fault system in the Kerby quadrangle—cuts the rocks between Monkey Creek and Monkey Creek Ridge; it is traceable northward from a point about half a mile north of Washington Flat to the point where it intersects the quadrangle boundary one mile west of the northeast corner. The course of the fault is marked by a series of saddles alined along spur ridges of Monkey Creek Ridge. Through the southern part of its course the fault originally thrust metavolcanic rocks of the Galice over slates of the Galice, but a later intrusion of gabbro along the fault has destroyed part of these relations. Near the north boundary of the quadrangle both sides of the fault are in peridotite and serpentine. Apparently there has been repeated movement along the fault, for a shear zone is exposed in the gabbro near the head of Monkey Creek, a faulted sliver of serpentine in the gabbro is exposed a short distance south of this shear zone, and gabbro is faulted against a tongue of peridotite north of the shear zone. The displacement along this fault is at least several thousand feet.

Parallel to the fault just described and about one mile west of it, in the vicinity of Baker Flat and Monumental, is a fault along the contact between gabbro to the east and slate to the west. Northward, where the fault cuts peridotite, a zone of slickentite has been produced. Small lenses of serpentine possibly sheared into place mark the trace of the fault on the hilltop east of Baker Flat. Other thin lenses of serpentine both in the gabbro and in the slate are believed to mark the traces of other similar, but smaller, faults.

The northward-trending diorite dike that cuts the peridotite in the vicinity of High Plateau Mountain and Diamond Creek ends southward in a zone of highly sheared serpentine. The dike and the sheared serpentine probably mark the course of another large reverse fault.

Later, at least in part, than the faults just described is a system of high-angle northwestward-trending faults. Three of these faults of considerable displacement were mapped in the upper reaches of Shelley and Monkey Creeks. The two northerly faults are mineralized in places, especially at the Blue Rock mine about half a mile southeast of Baker Flat. The mineralization at the Continental mine a short distance northwest of Monumental is probably controlled by the nearness of the northernmost fault. A fourth fault cuts off the peridotite and metavolcanic rocks about one mile south of the place where Siskiyou Fork enters the eastern edge of the quadrangle.

The third fault system, the Smith River system, consists of a large fault, two or three closely associated branch faults, and possibly the fault cutting serpentine and gabbro at the head of Diamond Creek. The large fault and its branches parallel the general course of Smith River between Idlewild Resort and the mouth of Patrick Creek. From the mouth of Patrick Creek the fault extends southwestward across, and offsets, the peridotite mass of Josephine Mountain. The fault cannot be traced definitely beyond the point where it offsets the west contact of the peridotite, but the gabbro exposed in Craigs Creek below French Hill may have been intruded along the continuation of the fault. This fault and its branches offset the folds, northward-trending reverse faults, and intrusive rocks, including a small body of gabbro exposed between Dollar Bend and Washington Flat. An apparent strike-slip movement of more than a mile is evident between Dead Horse Gulch and the mouth of Monkey Creek, but the magnitude of this horizontal component decreases southwestward; however, the rapidity with which the apparent horizontal displacement dies out southwestward and the sinuosity of the trace of the fault zone indicate that probably the displacement is largely dip-slip and shows a decided hingelike movement decreasing southwestward. The total displacement in the vicinity of Monkey Creek may be as much as 10,000 feet. The age relationship of this fault system to those trending northwestward is unknown, inasmuch as they were not observed to cross; both systems, however, are known to be younger than the ultramafic intrusive rocks.

STRUCTURAL FEATURES RELATED TO TERTIARY DEFORMATION

The rocks of the Gasquet quadrangle were probably deformed between Early Cretaceous time and late Miocene. The nature of the deformation is not decipherable, without much detailed work, within the limits of the quadrangle. The only structures readily manifest since Early Cretaceous time are those developed or developing since late Miocene.

After late Miocene, deformation in the Gasquet quadrangle has resulted in vertical uplift and slight tilting, but no folding and, so far as is known, no faulting have resulted from these disturbances. However, late Tertiary faults border the coastal plain a few miles west of the quadrangle. The land to the east of these faults has been lifted 1,500 to 2,200 feet above sea level, as shown by the presence of outcrops of the upper Miocene beds in the Gasquet quadrangle at altitudes of 2,000 feet and 2,200 feet. In addition to this uplift, a slight seaward tilt of about 2° has been imposed on the rocks in the Gasquet quadrangle. The uplift has been an intermittent, oscillatory process with periods of uplift and downdropping followed by relatively long interludes of quiescence. Recent uplift has rejuvenated many of the streams and caused them to incise their channels sharply to depths as great as 50 feet; terraces marks the old channels above present stream beds.

METAMORPHISM

Metamorphism of the rocks of the quadrangle may be considered in three categories: serpentinization, metamorphism induced by igneous instrusions, and metamorphism of dynamic origin. Serpentinization is restricted to the ultramafic rocks and is a complex process probably effected by more than one cause. Zones of metamorphism induced by igneous intrusion are not of wide extent and in comparison with the prevasive dynamic metamorphism are of little consequence.

All the ultramafic rocks in the quadrangle have been serpentinized to some extent. Because alteration to serpentine is ubiquitous, it seems probable that the greater part of the process was deuteric and that the water—water probably absorbed during earlier stages of the magmatic process—and the necessary silica stemmed directly from the peridotite itself during the final stages of cooling. However, there can be little doubt that serpentinization in some areas has been intensified by other processes. The intense alteration of the margins of ultramafic masses and shear zones may be due not only to shearing and a greater absorption of water, but also to localization and channeling of residual solutions. Diorite and gabbro intrusions may also affect the degree of serpentinization in adjacent ultramafic rock.

Metamorphism induced by igneous intrusions varies with the nature of the magma and of the invaded rock. Contact metamorphism effected by ultramafic intrusions is notably weak; the intruded rocks commonly show no change due to either heat or magmatic solutions, but shearing of the margins of the ultramafic rock and the adjacent host rock is both widespread and intense. The contact metamorphic effects of diorite and gabbro intrusions are much more pronounced. Slates have been indurated and locally altered to hornfels. Volcanic rocks have been intensely altered near contacts; the rocks have been thoroughly recrystallized, zones of igneous breccia are common, and considerable assimilation of country rock has occurred. However, less intense contact metamorphic effects remote from the igneous contacts have been masked by regional dynamic metamorphism.

All the pre-Tertiary rocks have been subjected to dynamic metamorphism. Intense pressure has folded and crushed the rocks and produced slaty cleavage in the finer grained sedimentary rocks. Concomitant with these physical changes, the nonultramafic rocks have undergone mineralogic transformations: feldspars have been saussuritized, and clay minerals have been converted to sericite, ferromagnesian minerals to chlorite and epidote, and augite to amphibole. These changes are characteristic of low-grade metamorphism, and the intense contact metamorphic effects remote from the igneous contacts mineral associations approach the "green schist" facies of Eskola (1921). Cataclastic textures in these rocks tend to indicate metamorphism was not deep-seated. Changes in the ultramafic rocks incident to dynamic metamorphism probably are largely mechanical. The highly altered slickentite of the shear zones is probably due to a combination of causes of which shearing, although the most effective, is only one.

GEOMORPHOLOGY

The geomorphic history of the Gasquet quadrangle as expressed by its topographic forms may possibly date back to Eocene time and certainly dates back as far as the Miocene, but insofar as these forms are controlled by the structures of the underlying rocks, the evolution of the present topography depends upon the entire geologic history of the quadrangle. The Gasquet quadrangle is a representative part of the Klamath Mountains and has essentially the same geomorphic history as the rest of the range, but some details of the physiographic development of the region are better preserved and are more easily interpreted in adjacent areas outside the quadrangle. Thus, the following discussion of geomorphic events involving the quadrangle is based in part on evidence that is well-preserved nearby but is either obscure or lacking within the quadrangle. Because the land forms have been dominantly controlled by late Tertiary events, it is justifiable to begin with a summary of probable conditions in late Miocene time.

THE KLAMATH PENEPLANE

By Miocene time a long period of erosion had reduced the Klamath Mountains to a surface of low relief and monotonous aspect. Remnants of this old surface are still preserved in the remarkably even, accordant ridge tops and in the broad, rolling upland surfaces exemplified in the Gasquet quadrangle by Pine Flat Mountain, High Plateau Mountain, Lower Coon Mountain, and others. Today, in fact, the Klamath Mountains, when viewed from a high point, give the impression of a dissected plateau (pl. 12) broken only by the group of rugged peaks to the east forming the crest of the Siskiyou Mountains and far to the south by another group, the Salmon Mountains. Ridge after ridge rises to the same general level until they appear to merge in the distance, where the mountainous aspect of the region is lost. This surface was called the Klamath peneplane by Diller (1902, pp. 15–18).

Bordering the western margin of this peneplane in and near the Gasquet quadrangle are scattered areas where marine upper Miocene beds crop out. The distribution of these sediments indicates an irregular coastline characterized by bays, islands, and promontories. Contrary to the views of earlier writers, however, the surface on which these beds were deposited is not continuous with the surface of the Klamath peneplane, nor were the beds laid down at the time when the peneplane had reached its greatest development. The surface on which the beds were deposited is separated from that of the peneplane by a distinct break in slope, and furthermore, the two surfaces slope seaward at different angles; the old peneplane surface about 2° and the surface on which the upper Miocene beds were deposited about 1°. In the vicinity of Bald Hill, 1½ miles west of the Gasquet quadrangle, the surface on which the upper Miocene beds were deposited is nearly 500 feet below the Klamath peneplane, but 6 miles east of Bald Hill, around the flanks of the central part of Lower Coon Mountain, this surface is fully 1,000 feet below the peneplane. (See pl. 14.)

Because the preserved record of the sequence of events is fragmentary, the relation of the upper Miocene beds to the development of the Klamath peneplane is not altogether clear; however, because of the great difference in altitude between the upper Miocene depositional surface and the peneplane, it seems improbable that the upper Miocene beds were deposited, as Diller thought, at the time of maximum development of the Klamath peneplane. The present writers believe the Klamath cycle of peneplanation ended with moderate regional uplift before the upper Miocene beds were deposited, and that with this uplift a new stage of erosion began. The Klamath peneplane was dissected and fairly wide valleys were carved into the old surface before the new cycle was partly halted by submergence. This submergence was not so great as the uplift that ended Klamath peneplanation but was sufficient to flood the lower ends of the river valleys. The upper Miocene beds were then deposited in the flooded river valleys and along the coast bordering the land mass. Haines Flat and the flat-topped summit of French Hill, both still partly covered by these beds, appear to be remnants of the floors of old valleys carved several hundred feet below the general level of the Klamath peneplane. The west end of Lower Coon Mountain projected, as an island, well above the level of the Wymer sea.

The difference in seaward slope between the Klamath peneplane and the surface on which the upper Miocene beds were deposited lends supporting evidence to the supposition that crustal movement intervened between Klamath peneplanation and deposition of the upper Miocene beds. This depositional surface probably had a slight seaward slope during late Miocene, but even if it were horizontal the seaward slope of the Klamath peneplane at the same time was approximately 1°, or about 90 feet per mile. This gradient seems excessive for an undisturbed old-age erosional surface approaching the degree of planation evident in the remnants of the Klamath peneplane. Therefore, in the opinion of the writers, not only uplift but also tilting, probably concomitant with uplift, occurred after the Klamath cycle of peneplanation and before deposition of the upper Miocene sediments.

LATER EROSION STAGES

Elsewhere in the Klamath Mountains and in the northern Coast Ranges of California, Diller (1902, pp. 11–15) recognized several stages of erosion younger than the Klamath peneplane. Similarly, several stages are recognizable in the Gasquet quadrangle, some of which correspond in general position to some of Diller's stages. As uplift in one area many accompany subsidence in another, correlation of stages between widely separated areas is difficult until detailed studies of intervening areas have been made. Because of the lack of sufficient data in areas lying between the type localities of Diller's stages and the Gasquet quadrangle, no attempt at correlations will be made.

Slight emergence followed deposition of the upper Miocene beds, the new deposits were trenched to some extent, and the gravels of French Hill, Lower Coon Mountain, and Haines Flat were deposited in the channels. Uplift could not have been very great in the area, because the gravels occur at essentially the same level as the upper Miocene beds. The coarseness of the gravels, however, suggests that greater uplift and more rapid erosion may have occurred farther east.

Little evidence remains by which the sequence of events immediately following deposition of the post-Miocene gravel can be established, but probably uplift was renewed. At any rate, it seems likely that the major streams had established themselves in essentially their present courses rather early in post-Miocene time and perhaps considerably earlier. Adjustments to bedrock of some of the secondary streams such as Monkey Creek, Shelley Creek, and Griffen Creek doubtless continued, but the forks of Smith River probably followed their present courses very closely. The Middle Fork of Smith River apparently adjusted its channel quite early to part of the Smith River fault zone; structural control of the course of the North Fork of Smith River is not apparent. The crest of Shelley Creek Ridge is considerably lower than Elk Camp Ridge and Monkey Creek Ridge on either side, and this lower elevation may indicate that early streams near base level wandered considerably before uplift renewed downcutting and the streams began carving their present valleys. Throughout most of its length the bed of Shelley Creek is several hundred feet above the beds of the neighboring parallel streams, Monkey Creek and the East Fork of Patrick Creek.

Later stillstands of base level have left plainly marked topographic evidence in the quadrangle. The most clearly defined of the later stillstands is that marked by a line of shoulders on spur ridges flanking the valleys of the Middle Fork of Smith River and its tributaries from Kelly Creek and above. In places these shoulders are prominent; elsewhere they have been obliterated by erosion. They occur between altitudes of 1,900 and 2,200 feet but are most prominent at about 2,100 feet. The Jones Creek that empties into the Middle Fork of Smith River appears to have entrenched an old-age valley of broad, gentle slope. This valley is probably about the same age as that represented by the shoulders or benches mentioned. Across the divide south of this valley the headwaters of Hurdygurdy Creek entrench an area of somewhat similar shape but at a higher altitude. Inasmuch as the valleys of both Jones and Hurdygurdy Creeks are in slate, it is probable that streams during earlier cycles were able to carve wider, relatively more mature valleys than streams in harder rock. Even today Jones Creek, in common with other larger creeks in slate belts, has a wider bed, lower gradient, and is nearer maturity than streams of comparable size in other rocks.

One striking feature about the remnants of the older valleys of the Middle Fork and other streams is the relatively constant altitude of these remnants from place to place; that is, they have a lower seaward gradient than the bed gradients of the present streams. The relatively flat gradient indicates the streams in the old valley were near base level and that the uplift that caused incision did not involve much seaward tilting in this area. Topographic features left by this stillstand of erosion, and for that matter, of other stillstands, are scarce in the part of this quadrangle south of the ridge separating the drainage of the Middle Fork of Smith River from that of the South Fork. The lack of these features is probably due to the fact that, by and large, only the headwaters of streams tributary to the South Fork of Smith River traverse the quadrangle, and, commonly, the topographic forms left by cycles of erosion are less well defined at the heads of streams because erosion is not only less rapid but is also steadier and less subject to fluctuations of land elevation.

Another period of relative land stability is probably indicated by the benches and shoulders at altitudes of 1,500 to 1,600 feet between Dead Horse Gulch and about a mile above the mouth of Siskiyou Fork. The gently sloping bench on the north side of Siskiyou Fork a short distance above its mouth and the low ridge between Washington Flat and Monkey Creek are deeply covered with old soils. Scattered benches and shoulders elsewhere and at different altitudes may point to other minor erosional pauses, but their correlation or interrelations are vague.

QUATERNARY EROSION SURFACES

The courses of both the Middle and South Forks of Smith River are market by gravel-covered terraces, many of them fully 75 feet above the present stream beds. The gravel on some of these terraces probably correlates with some of the marine terraces evident along the coast, which have been studied in detail by Griggs (1945) in the Coos Bay area in Oregon. The slightly incised valley floor of the Middle Fork of Smith River is nearly half a mile wide in the vicinity of Gasquet and is partly covered by terrace deposits. Elsewhere the valley floor is much narrower.

The present streams have entrenched these terraces as much as 75 feet, and in most places both the Middle and South Forks of Smith River are flowing on bedrock. Tributaries, notably those of the Middle Fork of Smith River above Kelly Creek, have incised the lower parts of their channels to form narrow box canyons a few tens of feet deep. The beds of the upper parts of these tributaries are gravel-filled over considerable distances, and apparently rejuvenation has not yet affected the valleys of these streams for more than short distances above their outlets.

There is little evidence of any piracy of large stream in the Gasquet quadrangle within the determinable past, but the captures of some large streams are now developing. Shelley Creek and Gordon

Creek appear to be the most imminent victims of beheading. About 2 miles up the East Fork of Patrick Creek from its junction with the West Fork a low ridge less than a quarter of a mile wide separates the East Fork from Shelley Creek, and the bed of the East Fork at this point is nearly 150 feet lower than the bed of Shelley Creek. Likewise, south of Baker Flat, Monkey Creek may capture the upper part of Shelley Creek; the bed of Monkey Creek is nearly 500 feet lower than that of Shelley Creek in the same area. Gordon Creek shows about the same relation to Coon Creek as Shelley Creek does to Monkey Creek, and gullies tributary to Coon Creek may eventually capture Gordon Creek north of Haines Flat.

LANDSLIDES

Landslides are a conspicuous and widespread topographic feature of the peridotite areas. Erosion has been rapid in the Gasquet quadrangle, and canyon walls are oversteepened beyond the point where peridotite, a structurally weak rock, can withstand gravitational stresses. Ground water undoubtedly aids landsliding by lubricating fissures along which sliding occurs. Peridotite is highly fractured and fissured, and the resulting permeability permits deep penetration of rain and snow water; thus, sliding is more active during the wet season. Landslides form irregular benches at various altitudes along the canyons that cut the peridotite. These benches, unlike stream terraces, slope toward the mountain rather than toward the streams. Depressions are common at the back of landslide benches, and the deeper depressions may contain ponds during the wet season. Many of the slides are very recent, and the grayish-green largely unweathered scars where the rock has slid away are visible from long distances. The surfaces of many of the more recent slides, especially the smaller ones, may be very rough and consist of jumbled rock. Beneath the surface the rock may be thoroughly broken, and open fractures are filled with mud and water.

Along the master streams cutting peridotite areas landsliding has largely obliterated traces of older erosion cycles. In fact, landsliding appears to be a major factor in erosion along the larger streams in peridotite areas. The canyon walls slide into the stream beds, and the streams cut away and remove the material. The water that seeps underground because of the high permeability of the peridotite later reaches the surface near the base of slopes and to some extent removes material at this level. Erosion by this process, called sapping, undoubtedly aids in oversteepening slopes by favoring more active erosion near the base of the slopes, and thus the oversteepened canyon walls become susceptible to renewed landsliding. As landslides change the surface distribution of some of the rocks and change their original relations, the large ones are shown on the map.

The visible record of the geologic history of the Gasquet quadrangle extends back no farther than perhaps Middle Jurassic time, and the record since then is by no means complete. No identifiable trace remains of some events that perhaps were as significant as those whose record is clear.

The quadrangle was part of a subsiding sea floor during Late Jurassic time; over this floor accumulated the thick deposits of sand, pyroclastic material, and lesser amounts of mud that make up the bulk of the Dothan formation. Some basaltic lava, including flows of pillow lava, was erupted. The massive, structureless character of the beds and the generally unweathered nature of the material suggest rapid deposition of material from high source areas undergoing rapid erosion.

Toward the end of Dothan time there was a marked increase in volcanic activity; great piles of lava and pyroclastic material forming the lower part of the Galice formation accumulated. Simultaneously, clastic sediments were being deposited, but these were commonly masked by the volcanic accumulations, except during short periods of eruptive quiescence. As the volcanic material accumulated the sea floor sank, but probably for short periods some areas were above sea level. The rock sequence, both within the quadrangle and along the strike to the north and south, suggests that an archipelago of low-lying volcanic islands existed in this region during the period of eruptive activity. Volcanism was predominantly explosive, and islands either rose above or fell below sea level according to whether volcanism or erosion—largely wave planation—dominated. The conglomerate facies of the volcanic breccia near Hurdygurdy Butte may indicate that parts of the area were directly connected to continental land masses but may merely indicate local shoaling which gave rise to near-shore types of deposits. The supposition that most of the material was deposited under water, however, is supported by the existence of pillow lava, and the fine-grained texture of the scattered slate lenses is suggestive of fairly deep water.

As the intensity of the volcanic activity subsided sediments now represented by the slate of the Galice formation were deposited, but deposition was interrupted at intervals by lava flows and ash showers. The presence of volcanic material throughout the sedimentary rock unit of the Galice probably indicates continued volcanic activity in the bordering land masses long after this activity had practically ceased in areas covered by the sea.

Events following the deposition of the Galice formation are obscure. Cretaceous rocks are nonexistent in the quadrangle, but rocks of the Horsetown formation of Cretaceous age crop out 5 or 6 miles northeast of the quadrangle near Waldo, Oreg. The nearest known Upper Jurassic Knoxville rocks are at least 40 miles from the quadrangle, but possibly both formations formerly covered the Gasquet quadrangle. Where Dothan, Galice, and Knoxville rocks occur in the same area, the Galice and Dothan rocks are deformed considerably more than those of the Knoxville, and it is reasonable to suppose that pre-Knoxville rocks were deformed in the Gasquet quadrangle before deposition of the Knoxville formation.

Diller (Diller and Kay, 1924) shows peridotite cutting Knoxville rocks in the Riddle quadrangle, Oreg., north of Gasquet quadrangle, and peridotite cobbles have been found in the Horsetown formation near Waldo, Oreg.; therefore, the peridotite is probably of late Knoxville age.

After deposition of the Galice formation the area underwent a period of strong folding, faulting, and igneous intrusion. Powerful southeasterly stresses forced the rocks into tight folds overturned to the west, and as pressure continued the rocks broke, forming steeply dipping thrust faults. As the rocks were being folded, huge sheets of ultramafic rocks intruded them, in places conformable to the strata, in others along unconformities, and elsewhere as dikes. Intrusions of diorite and gabbro followed the ultramafic intrusions, but all these igneous rocks were involved, at least partly, in the deformation of the stratified rocks.

Rapid erosion followed the period of diastrophism; the Horsetown formation near Waldo unconformably overlies the steeply dipping Galice formation. Little direct evidence of the former extent of the Upper Cretaceous Chico formation remains in the Klamath Mountains, but many geologists, including Diller, believe that all the area except a central island, now represented by the high central core of the Klamath Mountains, was formerly blanketed by the formation.

MINERAL RESOURCES

The mineral resources of the Gasquet quadrangle of proved or potential value include chromite, gold, mercury, asbestos, copper, and platinum. Of these, the value of the chromite produced far overshadows the combined value of the other mineral products.

HISTORY

Placer gold was discovered in Del Norte County in 1851. Copper was discovered in 1853 (Maxson, 1933, p. 145) and chromite about the same time (Maxson, 1933, p. 149). As the copper and chromite deposits were near the old Wymer stage road, and as the ore was useful to ballast vessels returning empty to the east coast, ores of copper and chromite were mined and shipped between 1860 and 1880.

Prior to 1880 nearly all mining, except placer, was in the area west of the Gasquet quadrangle, but in 1886 the French Hill chromite claims were patented. From the middle eighties until World War I exploration for copper and lode gold was carried on intermittently within the quadrangle. After the United States entered the war in 1917 prospecting for, and mining of, chromite was pursued vigorously until the collapse of the domestic chromite market following the armistice in 1918. Practically no mining was done during the next decade, but during the thirties a number of small placer mines were active. World War II stimulated chromite mining greatly; more than 20,000 long tons of chromite were shipped from the mines in the Gasquet quadrangle during this period.

Quicksilver deposits were discovered in the Diamond Creek area in the 1850's and were mined to some extent by the placer miners of the region for local use (Maxson, 1933, p. 150). The first recorded production of quicksilver in the quadrangle came from the Webb Cinnabar mine, which was discovered in October 1940, and began shipping quicksilver in 1943.

The gold produced from the Gasquet quadrangle during the period 1895–1930 was valued at $25,000; during the period 1931–40, 492.26 fine ounces was produced. These figures may be slightly in error, because it has been necessary to interpret the record and because some production may have come from parts of creeks outside the quadrangle. The production of gold prior to 1895 will never be known, but available information indicates that it was probably not greater than the production since 1895.

CHROMITE DEPOSITS

Thirty-three deposits of chromite have been mined in the Gasquet quadrangle and up to December 31, 1947, had yielded at least 30,000 long tons of lump ore. Although the precise value of the ore is not known, it probably exceeded $1,250,000. During World War II two of these mines, the French Hill and the High Plateau, were the largest producers of lump chromite in the country. Reserves amounting to many thousands of tons probably remain.

Chromite is usually present as an accessory mineral in all ultramafic rocks, but recoverable quantities are found only in dunite. Two types of primary chromite deposits, disseminated and pod, are found in the Pacific Coast States. Deposits of disseminated ore consist of grains and schlieren of chromite mixed with dunite, and pod deposits consist of irregular masses of nearly pure chromite.

In the Gasquet quadrangle only the pod deposits are of any value. The pods of chromite are enclosed in bodies of dunite, but the dunite may amount to only a thin rind. The weight of individual pods ranges from 1 to 10,000 long tons, but the majority weigh less than 100 long tons. So far as is known, chromite bodies are confined to dunite, which may occur anywhere within a peridotite mass. Pods of chromite are especially abundant in zones of shearing. It is not to be inferred, however, that shear zones localized the emplacement of chromite; instead, the reverse is true. Shears tend to seek out lines of least strength within a mass, and the contacts between chromite pods and the country rock are zones of weakness subject to shearing. Only those shear zones following preexisting pods of chromite contain ore.

The chromite deposits within the Gasquet quadrangle have been described in a recent bulletin (Wells, Cater, and Rynearson, 1946), and the reader is referred to this publication for a more complete discussion of the occurrences of chromite and a description of the deposits. New exploration has been done at the Holiday group of claims since the publication of this bulletin, and a brief description of this deposit follows.

HOLIDAY GROUP

The Holiday group of claims, formerly called the High Dome mine, is between the forks at the head of the West Fork of Patrick Creek in the NW¼, sec. 20, T. 18 N., R. 3 E. at an altitude of 2,500 feet. The property was located by Edward Cook who leased it to J. K. Remson and C. H. Bennett in 1942. During that year they mined and shipped 126 long tons of ore containing 44.6 percent Cr_2O_3 and having a Cr : Fe ratio of 2.48. The claims were abandoned in 1943 and were restaked by E. A. Carlson on July 1, 1948, as the Holiday claims.

The deposit is in peridotite a few thousand feet west of the eastern contact of the peridotite. The country rock is partly serpentinized saxonite cut by light-green veins of serpentine. The ore occurs as irregular tabular bodies of massive chromite in a zone of intensely sheared serpentine that strikes about N. 30° E. and dips 27° SE. Exposures were inadequate to determine the size and shape of the several pods, but the top surfaces of them were knobby. In places diamond drilling has penetrated as much as 10 feet of solid ore at right angles to the upper surface. Present indications suggest that this deposit probably contains several thousand long tons of lump ore.

GOLD DEPOSITS

Two types of gold deposits are found in the quadrangle—pyritic replacement bodies, veins, and fissures in pre-Tertiary rocks, and placer deposits in Recent and earlier gravels. The known veins are small and discontinuous. The mineralogy is simple: gold, pyrite, chalcopyrite, sphalerite, and in places a little galena, are associated in a gangue consisting largely of quartz with minor amounts of sericite and perhaps some carbonate. The placers have yielded most of the gold, which has been transported considerable distances and bears little direct relationship to the veins in the quadrangle. The placers have been worked intermittently for many years, but exploitation of the Miocene placer gravels has been severely handicapped by lack of water.

LODE DEPOSITS

Some lode deposits in the vicinity of Monumental have been prospected and mined. The deposits consist of replacement veins and irregular masses in gabbro, diorite, and metavolcanic rocks in the vicinity of strong northwestward-trending faults. The veins and mineralized zones consist of fine-grained quartz, sericite, pyrite, chalcopyrite, sphalerite, galena, and gold. They are commonly brecciated, and the brecciated zones are favored sites for the sulfides. Exploration has failed to uncover any strong persistent veins or irregular replacement deposits of large size.

CONTINENTAL MINE

The Continental mine is on a tributary of Shelley Creek about a quarter of a mile north-northwest of Monumental. Workings consist of two or three caved adits, several prospect pits, and an irregular adit having a short inclined shaft leading to a short lower level.

The country rock is a brecciated fine-grained meta-andesite flow locally called "greenstone." It is silicified in places and elsewhere has been completely replaced by quartz and sulfides, but no definite, well-defined vein exists. The brecciation appears to be related to the strong northwestward-trending fault that cuts the rocks a short distance north of the mine. The scattered prospect pits indicate that the mineralization is spotty. A carload of ore shipped to the smelter in 1946 brought $15.90 per ton. The deposits apparently have been worked intermittently for many years, and many of the old pits are now overgrown with brush. The early history of the mine is not known. In March 1946, the property was leased from William Stanfield by E. V. Cook and W. R. Smith, who were mining ore through the summer of 1946.

BLUE ROCK MINE

The Blue Rock mine is half a mile southeast of Baker Flat in the canyon of Monkey Creek. Workings consist of a few short adits. No information regarding ownership or history was obtained.

The country rock is sheared and brecciated gabbro and diorite cut by a number of small quartz veins and stringers, which are short, irregular, and discontinuous. Small amounts of pyrite are scattered through the quartz, and copper stains are common. Mineralization has been controlled by the strong northwestward-trending fault immediately west of the mine.

At the time of this survey all the placers in the Gasquet quadrangle were inactive. Records of production are inaccurate, but the size of the workings indicates that the gravels on the old erosion surface were probably the most productive; insufficient water has always been a handicap in exploiting them. No known systematic sampling has ever been done, and for this reason the value of the gravels is unknown. The quantity of recent gravels is not large, but usually sufficient water is available to operate the placer mines in them the year round.

FRENCH FLAT PLACERS

The placers on French Flat have been worked more extensively than any others in the quadrangle. They have been in intermittent operation since their discovery in 1860, and large reserves of gravel remain. A ditch several miles long was constructed in 1877 to bring water from the head of Coon Creek to wash the gravels on French Hill, but the Morrell placers a mile to the southwest depend on the scanty local flow.

The gravel that has been worked is 5 to 20 feet thick. Most of the gravels are coarse and contain boulders more than a foot in diameter; clay lenses are common. Much of the gold is relatively pure; 15 ounces from the French Hill placer mine in 1940 was 924 fine. Occasionally, nuggets worth as much as $15.00 are obtained from these placers.

LOWER COON MOUNTAIN PLACERS

Gravels at least 30 feet thick have been worked on Lower Coon Mountain. These are at an altitude of about 2,200 feet and appear to have been worked most extensively on the headwaters of side creeks tributary to Craigs Creek. The road on the east side of Lower Coon Mountain was built largely on the remains of an old placer ditch which carried water from Coon Creek to the placers, but undoubtedly operations were always handicapped by water shortages. The deposits are by no means as large as those of French Flat, or Haines Flat, nor have they been so extensively worked.

HAINES FLAT PLACERS

Tertiary gravels are abundant on Haines Flat; more of them are outside the quadrangle than within. These gravels have been worked at a number of localities since their discovery in 1877, but all the properties have been idle for many years. Remains of an old placer ditch are still present high on the east side of Coon Creek canyon; presumably this ditch formerly carried water to Haines Flat. Much gravel remains to be worked in this area, but nothing is known regarding the gold content.

Quaternary gravels have been worked at a number of localities on Monkey Creek, especially below the Blue Rock mine a mile southeast of Baker Flat and on the terraces along the lower 2 miles of the creek. The terraces along the lower part of the creek are 20 to 40 feet above the present stream which is incised in bedrock. Placer gravels are not abundant in the stream bed, and presumably the more productive stream gravels have been worked out. Considerable unworked gravel remains on the terraces.

SMITH RIVER PLACERS

Placer gravels along the Middle Fork of Smith River have been worked on a small scale, but results have not been encouraging. Much of the stream bed is solid rock almost devoid of gravel, and elsewhere the gravels are thin. Many of the terraces are covered with sand and gravel 10 to 20 feet thick, but these have been worked only in a few places.

CRAIGS CREEK PLACERS

The Recent gravels that have been most extensively worked are those in Craigs Creek. Possibly the richness of these deposits may be due to a concentration of gold washed from the old Tertiary gravels on the ridges and flats above the creek. For a distance of 2 miles east of the place where Craigs Creek leaves the quadrangle, the stream and bank gravels have been worked extensively wherever sizable accumulations of material have been found. Some of the placers were intermittently active until about 1940.

COPPER DEPOSITS

THE CLEOPATRA MINE

The only copper deposits of any value are on or just below the old upland surface in the northwest corner of the quadrangle. These deposits resulted from the leaching and secondary enrichment of very low grade pyritic ores. Downward-percolating surface waters containing dissolved air oxidized the sulfides; most of the copper was carried downward to the water table and deposited as new sulfide compounds, and where the process was continued long enough, native copper was formed.

The most important of these deposits, worked at the Cleopatra mine, is in the extreme northwestern corner of sec. 3, T. 18 N., R. 2 E. at an altitude of 2,350 feet. The deposit is in the peridotite mass of Josephine Mountain 2,000 feet from the contact with the Dothan formation and is therefore in the lower part of the mass. Rhyolite dikes trending slightly west of north are common in the vicinity of the mine. When the mine was visited all the workings were caved, and nothing could be learned about the character of the ore body. The late John Harvey of Grants Pass, Oreg., had a piece of native copper weighing about 150 pounds, which, according to him, was representative of the ore shipped from the property. Enclosed within the copper were small lumps of serpentine. This deposit is probably similar to the deposits on Copper Creek 5 miles to the southwest. These also are near the base of the peridotite mass of Josephine Mountain and consist of bodies of massive sulfides.

The copper content of the primary sulfide is less than 0.05 percent, but where these deposits have been exposed to oxidation and downward enrichment of the copper, usable ore has been formed. At depth these deposits grade into massive sulfide containing little or no copper. The Cleopatra mine is just below the surface of the Klamath peneplane, and therefore the deposit has been subjected to secondary enrichment for a prolonged period. It is reasonable to deduce that the deposit of native copper graded downward into unenriched sulfide. A carload of native copper is said to have been shipped from the property during the latter part of the last century.

QUICKSILVER DEPOSITS

Cinnabar and native quicksilver deposits have been found in the Patrick Creek and Diamond Creek areas. The Patrick Creek deposit is in peridotite within a few thousand feet of its eastern contact and is associated with Tertiary(?) intrusive rocks, whereas the Diamond Creek deposits occur either in peridotite where intruded by diorite or in the altered diorite itself. In the Diamond Creek area the peridotite is completely altered to serpentine and is highly sheared along the diorite contact. In 1946, workings in the Diamond Creek area were so caved and covered that little could be seen, and therefore the description of the deposits which follows is taken from Maxson's report (1933, pp. 156–158) with but slight changes and omissions.

DIAMOND CREEK DEPOSITS

Cinnabar was discovered on Diamond Creek in the NW¼ of sec. 11, T. 18 N., R. 2 E. in the early fifties. At that time placer miners from as far as Kerby, Oreg., a distance of 35 miles, went there to obtain quicksilver for amalgamation processes. K. J. Khoeery, reports on the authority of Bill Wimer, that a shaft was sunk on the cinnabar ledge. A fire was built in the shaft, and quicksilver condensed on the walls, which were wetted occasionally to keep them cool. After this crude roasting the quicksilver was collected from the bottom of the shaft. The property was first located by an English company in the sixties and was claimed by many other people thereafter. John Taggart and associates of Smith River claimed the property in 1916 and in the following year built a retort having three units and underdraft. The 37-inch pipe condensers ran into water. The retort had a capacity of 500 pounds of ore and was retorted for 6 hours. One flask of quicksilver was recovered in 1917, and operations ceased when it was found that most of the quicksilver was being lost because of the faulty retort.

SUNNY BROOK PROSPECT

This property was held in 1931 by Lee Brown of Los Angeles, who has two center claims, and by John Taggart of Smith River, who has two claims to the south and one to the north. In 1930 some development work was done, and the main shaft, which was caved in 1917, was uncovered. A crosscut below the creek level was driven to connect with an old tunnel in which pockets of native quicksilver had been found during earlier exploration.

The workings and retort are on an extensive terrace on the northwest side of Diamond Creek. The ore-bearing ledge is covered by a few feet of creek boulders. A northward-trending cut 40 feet long and 10 feet wide lies 100 feet north of the retort.

The ore mineral, cinnabar, occurs as stringers in quartz veins which cut the serpentine country rock and in fissures in the serpentine itself. There seem to be two highly fissured parallel lodes striking due north, which may be associated with the broad dikes of hornblende diorite outcropping 100 feet to the northwest.

Some of the ore was good grade and contained 1 to 2 percent mercury, but the average tenor was reported to be 10 pounds of quicksilver per ton.

BIG BOY CINNABAR GROUP

The Big Boy cinnabar deposits are on the north fork of Diamond Creek, at an altitude of 2,150 feet, about 4 miles northeast of the Sunny Brook group. Two claims in Curry County, Oreg., are owned by R. E. Strayner and J. J. Hoogstraat. Three claims in California are owned by O. H. Hagberg, H. W. Lipple, and George Davis. The group was operated under a partnership arrangement.

The cinnabar is scattered along fine joint fissures in a mass of propylitized diorite. The feldspar in this rock has been completely altered to kaolinite and sericite, and the amphibole has been altered to limonite. The altered diorite is exposed over the top of the ridge west of the camp and farther west is in contact with, or continuous with, the dikes passing near the Sunny Brook prospect. A tongue of serpentine crops out to the north, and to the east less altered rocks occur, including either a fresh hornblendite or gabbro containing inclusions of serpentine.

The original locator, John Griffin, dug a ditch along the top of the ridge and ground-sluiced what was originally a small slide. The water was run through a 10-inch sluice box equipped with Hungarian block riffles. The concentrates were retorted in two 4-inch pipes. Later equipment, installed by the J. I. L. Dredging Co. of Spokane, which leased the property, was essentially a refinement of the above. A 3-inch giant was operated in the slide and the material run through a series of sluices in an attempt to concentrate the heavier cinnabar crystals by gravity separation. The process was extremely inefficient, operations were abandoned, and the property has been idle a number of years.

WEBB CINNABAR MINE

Location.—The Webb Cinnabar mine comprises a group of nine claims on the west slope above the West Fork of Patrick Creek. The property is accessible from U. S. Highway 199 over 6 miles of dirt road, but the last 3 miles of this road is impassable during the long periods of wet weather.

History and production.—In 1948 the mine was called the Shults mine after its owners, M. D. Shults and I. N. Shults. As the mine has had many names during its brief existence, it seems advisable to reinstate the original name, the Webb Cinnabar mine, given by the discoverer, David L. Webb. Mr. Webb panned some cinnabar while prospecting Patrick Creek in August 1940. He traced the cinnabar upstream and out onto the hillside after no more of the mineral was found in the stream bed. As the ore did not crop out, the location spot was found entirely by panning, which is probably the oldest and most rewarding of geophysical methods. A discovery trench and an adit were started by Webb in October 1940. Later, the claim was sold to H. C. Wilmot, who thereafter located eight other claims. Prospecting by trenching was continued, and a road was built to the property in the autumn of 1941.

Exploration and mining were done intermittently during 1942 and 1943 by Burton B. Avery who subleased the property, installed two 12-inch steel pipe retorts 7 feet long, and produced 15 flasks of quicksilver in 1943. Wilmot relinquished his interest in the property in 1944, and the property was acquired by Oscar E. Hanno, who built a retort and shipped 7 flasks of quicksilver in 1944. In 1945, 68 flasks were produced from 103 tons of ore, a yield of 45.5 pounds of quicksilver per ton. The Shultses, who bought the mine in October 1946, constructed a rotary kiln 30 inches in diameter and 30 feet long, having a rated capacity of 30 tons in 24 hours. They continued exploring the property with tunnels.

Workings.—By September 1948, workings consisted of two adits and numerous prospect trenches and open-cuts. The total length of all the underground workings was about 450 feet. (See pl. 15.)

Geology.—The mineralized area is about 3,000 feet west of the contact of the peridotite and metavolcanic rocks of the Galice formation. This contact, like many other contacts of peridotite masses, is highly sheared. Quartz diorite has been intruded along it, but the intrusions are too small to be shown on the quadrangle map. Quicksilver-bearing rocks occupy an eastward-trending area about 800 feet long and 600 feet wide on both sides of Cinnabar Gulch. Webb reported that he panned some cinnabar from the next gulch north but found none in place.

The country rock is serpentinized saxonite, the degree of serpentinization ranging from 25 to 100 percent. The highly serpentinized rock is strongly sheared, soft, weathers rapidly, and generally fails to crop out; the fresher, less serpentinized saxonite is unsheared, resists weathering, and crops out in bold ledges. The fresher saxonite is jointed in a reticulate pattern, which cuts the rock into blocks measuring 1 to several feet on a side.

The sheared serpentine is cut by numerous small irregular intrusions that range from 1 to 100 feet in maximum exposed dimensions. Postintrusive shearing may have accentuated the irregularity of these bodies. The rock types, on the basis of field examinations, were grouped as felsite, andesite, basalt, and diorite. The felsite is aphanitic, light gray to light brown, and strongly iron-stained along fractures. Many dikes are highly kaolinized. One outcrop of felsite, showing a fine banding resembling flow structure, is associated with porphyritic andesite and appears to be a contact phase of that rock, although the actual contact with serpentine is covered. In general the felsite masses have an easterly trend. Two kinds of porphyritic andesite occur: one with feldspar phenocrysts, the other with hornblende phenocrysts. Visible quartz is absent. The hornblende andesite is fresh, but the andesite with feldspar pheoncrysts is commonly somewhat kaolinized. A number of outcrops of a dark aphanitic nonporphyritic rock containing a few small vesicles are thought to be basalt. A few dikes of a fine- to medium-grained gray to medium-brown rock were identified in the field as diorite. The rock contained feldspar and hornblende but no visible quartz. No evidence of the age relationships of these postperidotite intrusives were seen in the mined area, but from evidence elsewhere in the quadrangle (see p. 105) probably the felsite, and perhaps the basalt, is much younger than the diorite and andesite.

Because of the difference in structural competence between the masses of hard dike rocks, the less resistant blocks of relatively fresh saxonite, and the very weak, almost plastic, serpentine (which is, in a sense, a matrix enclosing the harder rocks), structures are neither systematic nor continuous. The fracturing and faulting with which mineralization is associated is best exposed in the main tunnel of the mine. At this locality are two major sets of fracturing: one trends about N. 70° W. and dips steeply north; the other trends about N. 20° E. and dips 40° to 60° E. During early shearing the first set may have splintered out into the second set. However this may be, the ore deposits occur as replacements, most commonly at the intersections of the two major sets of fracturing, and tend to occur intermittently and diminishingly along the northwestward-trending set. In general deposits are larger and richer in areas where the fractures cut felsite masses.

In the ore-bearing zones of fractures the dike rocks, especially felsite, have been altered to kaolin, and the serpentine to silica-carbonate rock. Apparently kaolinization occurred early in the ore-forming epoch, for quicksilver is found in the clay. Abundant silica introduced early in the period of mineralization silicified the serpentine and formed silica-carbonate rock, a rock characteristic of quick-silver deposits in serpentine throughout the Coast Ranges in California. The character of the silica-carbonate rock in the mine area varies with the degree of alteration of the serpentine, ranging from a dark-green fine-grained rock preserving the texture of the serpentine to a light-gray or white hard homogeneous rock that breaks with a blocky fracture. The highly altered rock may easily be mistaken for altered felsite, but commonly close examination reveals unreplaced grains of chromite and magnetite. Thin carbonate veinlets are numerous.

Cinnabar, pyrite, and possibly metacinnabar are the only sulfides present. Finely crystalline cinnabar and pyrite occur in the silica-carbonate rock or in places as incrustations on rock fragments in brecciated areas; both minerals are also found in clay gouge. Considerable quantities of a black, apparently crystalline, mineral were found in the original adit and in an open-cut. A spot test indicated that at least some of it was a compound of mercury, and it was tentatively identified as metacinnabar. Native quicksilver was abundant in the original adit and in the first crosscut to the east in the main tunnel. The surface of the rock was covered with tiny globules of quicksilver, and the rock assayed as high as 5.00 percent quicksilver.

Most of the ore was taken from the main tunnel and in the original adit. The mineralized zone in these workings is sheared and brecciated; along some of the shears the gouge is a foot thick and contains silica-carbonate inclusions several inches across. Most of the rocks in this zone are highly altered; silica-carbonate is abundant, and the felsite is kaolinized. The gouge, the inclusions in the gouge, and the surrounding rock all carry cinnabar and native quicksilver. The ore zone is cut off on the south by a postmineral fault that strikes N. 70° E.

to east and dips 60°–80° N. In adit no. 1 this fault seems to intersect, and then follow, the shallower-dipping mineralized zone of fracturing.

The ore zone explored in the adit and the main tunnel has been explored on the surface by open-cuts for a distance of 80 feet westward from the adit. Another cut 210 feet west of the adit exposes ore, but it is not known whether this is part of the same ore zone.

Since the classic work of Becker (1888), quicksilver deposits have been considered as being deposited from hot waters derived from cooling igneous rocks. Lindgren (1928) classifies them as "epithermal deposits" and indicates that they are usually associated with Tertiary and Recent lavas. None of these lavas, if ever present, remain in or near the Gasquet quadrangle, but the felsite dikes, believed to be of Tertiary age, could readily have been related to mineralization.

In the Pacific Coast states economically noteworthy deposits of quicksilver are found where brecciated rock masses offering a large amount of surface per unit volume are within 1,000 or 2,000 feet of the surface and are susceptible to mineralization. If a large deposit is to be formed, these rock masses also must be of large size and must be acted upon by large volumes of quicksilver-bearing solutions over a sufficiently long period of time. Most of these physical conditions are present in the Patrick Creek area, but whether enough quicksilver-bearing solutions were present to make large deposits is not known. The absence of large amounts of silica-carbonate rock in the area is inconclusive evidence that no large deposits were formed.

ASBESTOS

Small amounts of amphibole asbestos were mined during World War II from veins and seams in serpentine at scattered localities in the quadrangle. No single deposit yielded more than a few pounds, and it seems unlikely that any large deposits exist.

NICKEL

Many peridotite bodies contain small amounts of nickel. It occurs in the olivine, the nickel content of which may be as much as one-third of 1 percent but is commonly much less. Olivine breaks down upon weathering and liberates silica, magnesium, iron, and the nickel that is present. The nickel may be taken out of solution either by being absorbed by the iron hydroxide which precipitates out of the ground water when the contained ferrous salts hydrolize or by forming the characteristically green nickel silicate, garnierite. Thus, under favorable conditions of weathering, an ore deposit may be formed on and below an erosion surface. Garnierite has been found in the peridotite along the edge of the Illinois River Valley in the Kerby quadrangle, and a deposit has been described by Pecora and Hobbs (1942) near Riddle, Oreg., about 50 miles north of the Gasquet quadrangle. It is possible that similar deposits may be found in the Gasquet quadrangle.

PLATINUM AND ALLIED METALS

Small amounts of the platinum group metals have been recovered from the placer deposits along Craigs Creek. Little is known concerning the composition of this "platinum," but it is probably similar to that found elsewhere in the region and consists of an alloy of platinum, iridium, osmium, and ruthenium. These metals are undoubtedly derived from the peridotite.

LITERATURE CITED

Becker, G. F., 1888, Geology of the quicksilver deposits of the Pacific slope: U. S. Geol. Survey Mon. 13.

Bowen, N. L., and Schairer, J. F., 1936, The problem of the intrusion of dunite in the light of the olivine diagrams: Rept. 16th Internat. Geol. Cong., Washington, 1933, vol. 1.

Butler, G. M., and Mitchell, G. J., 1916, Preliminary survey of the geology and mineral resources of Curry County, Oregon: Oregon Bur. Mines, Min. Res. Oregon no. 2.

Cloos, Ernst, 1947, Boudinage: Am. Geophys. Union Trans., vol. 28, no. 4, pp. 626, 632.

Dewey, H., and Flett, J. S., 1911, On some British pillow lavas and the rocks associated with them: Geol. Mag., vol. 8, pp. 204–205.

Diller, J. S., 1902, Topographic development of the Klamath Mountains: U. S. Geol. Survey Bull. 196.

——— 1914, Mineral resources of southwestern Oregon: U. S. Geol. Survey Bull. 546.

Diller, J. S., and Kay, G. F., 1924, Description of the Riddle quadrangle [Oregon]. U. S. Geol. Survey Geol. Atlas, folio 218.

Eskola, Pentti, 1921, The mineral facies of rocks: Norsk geol. tidsskr., vol. 6, p. 143.

Gilluly, James, 1935, Keratophyres of eastern Oregon and the spilite problem: Am. Jour. Sci., 5th ser., vol. 29, pp. 225–252, 336–352.

Griggs, A. B., 1945, Chromite-bearing sands of the southern part of the coast of Oregon: U. S. Geol. Survey Bull. 945–E, pp. 113–150. [1946.]

Hershey, O. H., 1911, Del Norte County (Calif.), geology: Min. and Sci. Press, vol. 102, p. 468.

Lindgren, Waldemar, 1928, Mineral Deposits, pp. 538–541, New York, McGraw-Hill Book Co., Inc.

McKay, Robert A., 1946, The control of impounding structures on ore deposition: Econ. Geology, vol. 41, pp. 13–46.

Maxson, J. H., 1933, Economic geology of portions of Del Norte and Siskiyou Counties, northwesternmost California: California Jour. Mines and Geology, vol. 29, nos. 1 and 2, pp. 123–160.

Park, C. F., Jr., 1946, The spilite and manganese problems of the Olympic Peninsula, Washington: Am. Jour. Sci., vol. 244, pp. 305–323.

Pecora, W. T., and Hobbs, S. W., 1942, Nickel deposit near Riddle, Douglas County, Oregon: U. S. Geol. Survey Bull. 931–I, pp. 205–226.

Sosman, R. B., 1950, Centripetal genesis of magmatic ore deposits (abstract): Geol. Soc. America Bull., vol. 61, pt. 2, p. 1505.

Taliaferro, N. L., 1942, Geologic history and correlation of the Jurassic of southwestern Oregon and California: Geol. Soc. America Bull., vol. 53, pp. 71–112.

U. S. Dept. of Agriculture, 1926, Weather Bureau summaries, Climatological data by sections, Bull. W Section Sixteen, p. 11.

Wells, F. G., Cater, F. W., Jr., and Rynearson, G. A., 1946, Chromite deposits of Del Norte County, California: California Div. Mines Bull. 134, part 1, chapt. 1.

Wells, F. G., and Walker, G. W., Geology of the Galice quadrangle, Oregon: U. S. Geol. Survey, Geol. Quad. Map.   (In press.)

Wells, F. G., and others, 1949, Preliminary description of the geology of the Kerby quadrangle, Oregon: Oregon Dept. Geol. and Min. Industries Bull. 40.

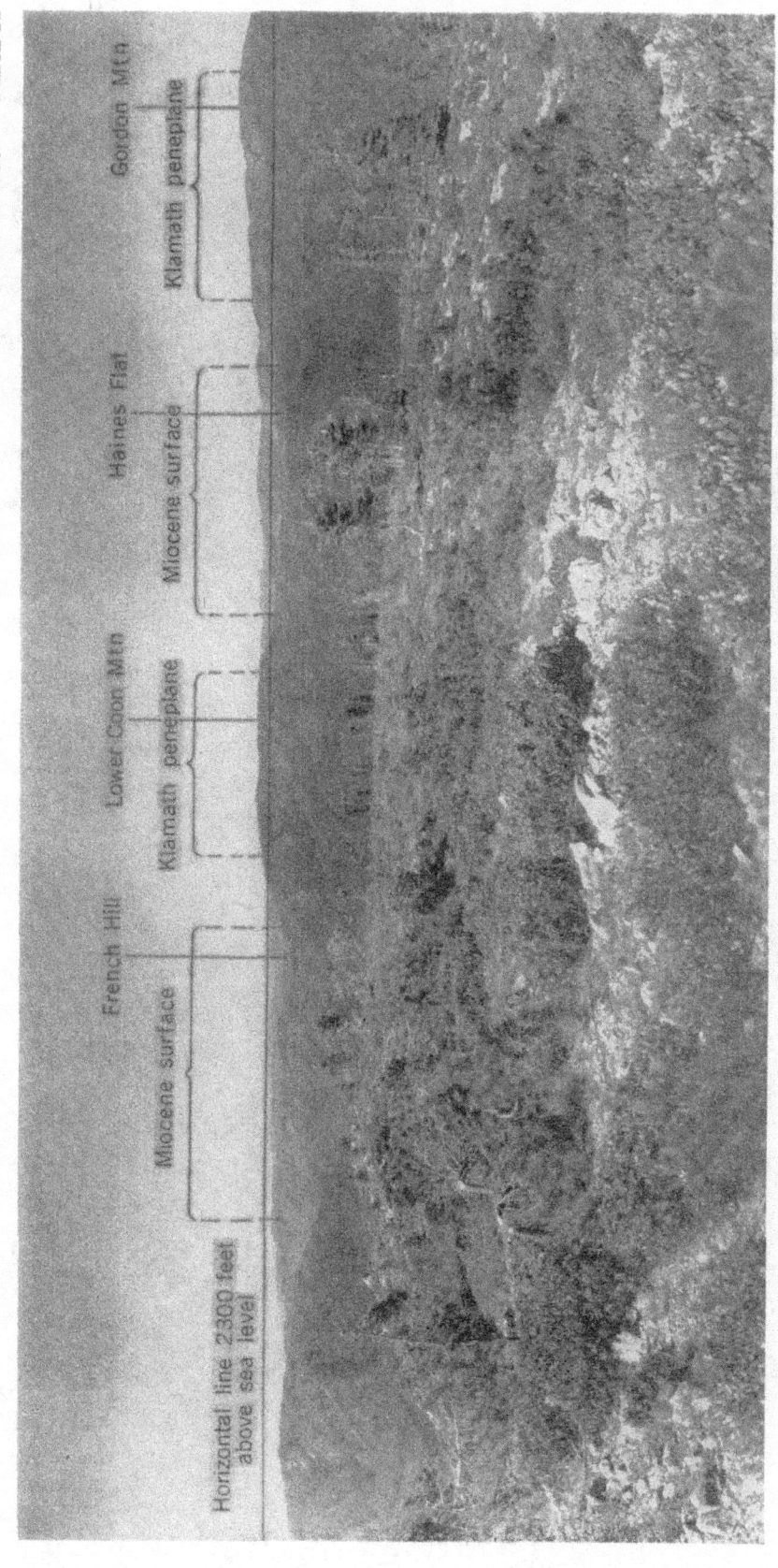

PANORAMIC VIEW OF THE KLAMATH MOUNTAINS

View northward from the Forest Service lookout on Rattlesnake Point, 4½ miles south-southeast of the mouth of Coon Creek.

A. A CLOT OF DUNITE IN SAXONITE

View of outcrop on north end of Elk Camp Ridge, showing the smooth surface of weathered dunite and
rough surface of weathered saxonite. Note veinlets of serpentine.

B. TERRACE DEPOSITS ALONG SMITH RIVER

View of deposit exposed on north bank of the river, about 1 mile west of Gasquet quadrangle.

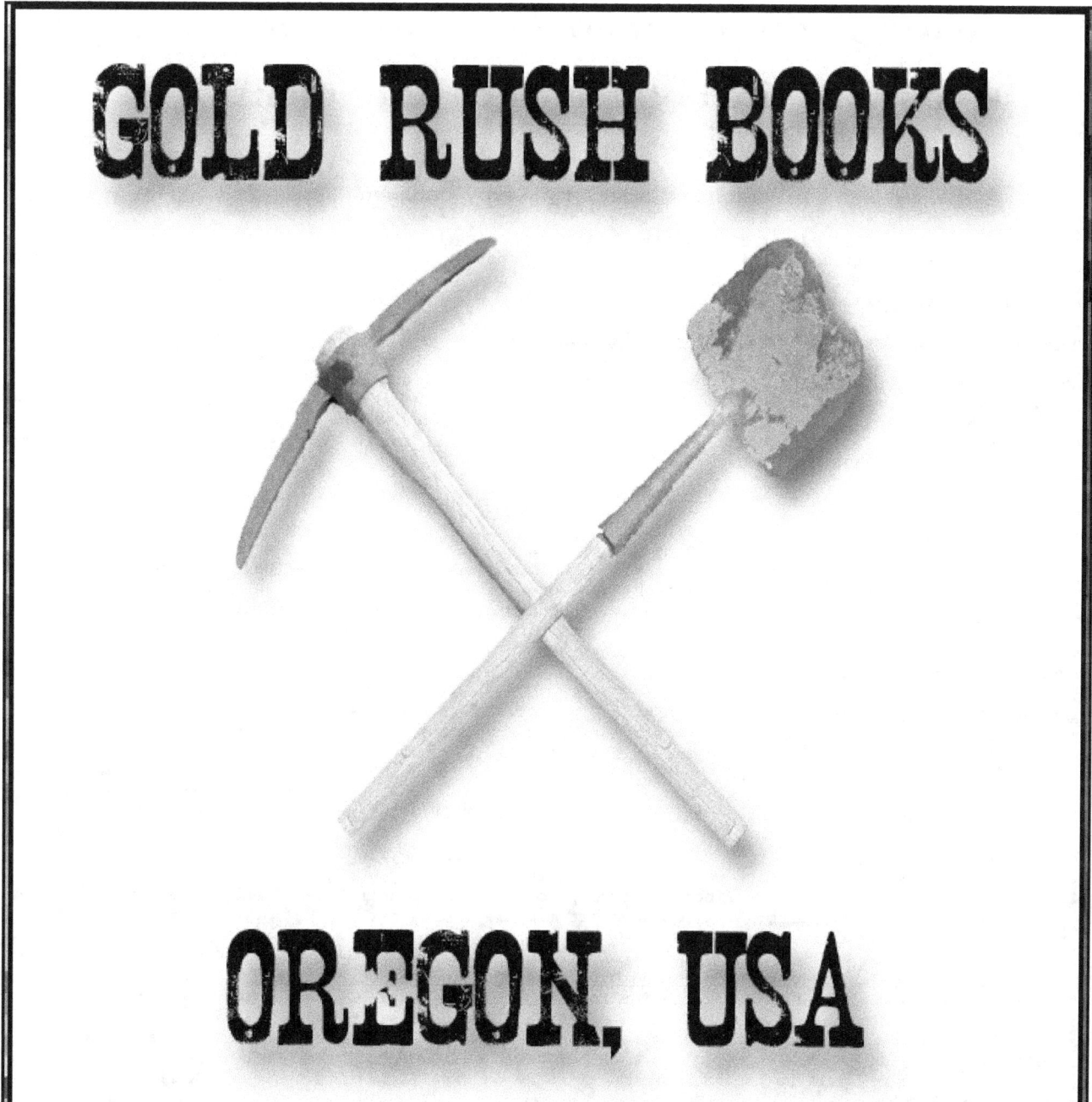

GOLD RUSH BOOKS

OREGON, USA

www.GoldMiningBooks.com

Books On Mining

Visit: www.goldminingbooks.com to order your copies or ask your favorite book seller to offer them.

Mining Books by Kerby Jackson

Gold Dust: Stories From Oregon's Mining Years - Oregon mining historian and prospector, Kerby Jackson, brings you a treasure trove of seventeen stories on Southern Oregon's rich history of gold prospecting, the prospectors and their discoveries, and the breathtaking areas they settled in and made homes. 5" X 8", 98 ppgs. Retail Price: $11.99

The Golden Trail: More Stories From Oregon's Mining Years - In his follow-up to "Gold Dust: Stories of Oregon's Mining Years", this time around, Jackson brings us twelve tales from Oregon's Gold Rush, including the story about the first gold strike on Canyon Creek in Grant County, about the old timers who found gold by the pail full at the Victor Mine near Galice, how Iradel Bray discovered a rich ledge of gold on the Coquille River during the height of the Rogue River War, a tale of two elderly miners on the hunt for a lost mine in the Cascade Mountains, details about the discovery of the famous Armstrong Nugget and others. 5" X 8", 70 ppgs. Retail Price: $10.99

Oregon Mining Books

Geology and Mineral Resources of Josephine County, Oregon - Unavailable since the 1970's, this important publication was originally compiled by the Oregon Department of Geology and Mineral Industries and includes important details on the economic geology and mineral resources of this important mining area in South Western Oregon. Included are notes on the history, geology and development of important mines, as well as insights into the mining of gold, copper, nickel, limestone, chromium and other minerals found in large quantities in Josephine County, Oregon. 8.5" X 11", 54 ppgs. Retail Price: $9.99

Mines and Prospects of the Mount Reuben Mining District - Unavailable since 1947, this important publication was originally compiled by geologist Elton Youngberg of the Oregon Department of Geology and Mineral Industries and includes detailed descriptions, histories and the geology of the Mount Reuben Mining District in Josephine County, Oregon. Included are notes on the history, geology, development and assay statistics, as well as underground maps of all the major mines and prospects in the vicinity of this much neglected mining district. 8.5" X 11", 48 ppgs. Retail Price: $9.99

The Granite Mining District - Notes on the history, geology and development of important mines in the well known Granite Mining District which is located in Grant County, Oregon. Some of the mines discussed include the Ajax, Blue Ribbon, Buffalo, Continental, Cougar-Independence, Magnolia, New York, Standard and the Tillicum. Also included are many rare maps pertaining to the mines in the area. 8.5" X 11", 48 ppgs. Retail Price: $9.99

Ore Deposits of the Takilma and Waldo Mining Districts of Josephine County, Oregon - The Waldo and Takilma mining districts are most notable for the fact that the earliest large scale mining of placer gold and copper in Oregon took place in these two areas. Included are details about some of the earliest large gold mines in the state such as the Llano de Oro, High Gravel, Cameron, Platerica, Deep Gravel and others, as well as copper mines such as the famous Queen of Bronze mine, the Waldo, Lily and Cowboy mines. This volume also includes six maps and 20 original illustrations. 8.5" X 11", 74 ppgs. Retail Price: $9.99

Metal Mines of Douglas, Coos and Curry Counties, Oregon - Oregon mining historian Kerby Jackson introduces us to a classic work on Oregon's mining history in this important re-issue of Bulletin 14C Volume 1, otherwise known as the Douglas, Coos & Curry Counties, Oregon Metal Mines Handbook. Unavailable since 1940, this important publication was originally compiled by the Oregon Department of Geology and Mineral Industries includes detailed descriptions, histories and the geology of over 250 metallic mineral mines and prospects in this rugged area of South West Oregon. 8.5" X 11", 158 ppgs. Retail Price: $19.99

Metal Mines of Jackson County, Oregon - Unavailable since 1943, this important publication was originally compiled by the Oregon Department of Geology and Mineral Industries includes detailed descriptions, histories and the geology of over 450 metallic mineral mines and prospects in Jackson County, Oregon. Included are such famous gold mining areas as Gold Hill, Jacksonville, Sterling and the Upper Applegate. 8.5" X 11", 220 ppgs. Retail Price: $24.99

Metal Mines of Josephine County, Oregon - Oregon mining historian Kerby Jackson introduces us to a classic work on Oregon's mining history in this important re-issue of Bulletin 14C, otherwise known as the Josephine County, Oregon Metal Mines Handbook. Unavailable since 1952, this important publication was originally compiled by the Oregon Department of Geology and Mineral Industries includes detailed descriptions, histories and the geology of over 500 metallic mineral mines and prospects in Josephine County, Oregon. 8.5" X 11", 250 ppgs. Retail Price: $24.99

Metal Mines of North East Oregon - Oregon mining historian Kerby Jackson introduces us to a classic work on Oregon's mining history in this important re-issue of Bulletin 14A and 14B, otherwise known as the North East Oregon Metal Mines Handbook. Unavailable since 1941, this important publication was originally compiled by the Oregon Department of Geology and Mineral Industries and includes detailed descriptions, histories and the geology of over 750 metallic mineral mines and prospects in North Eastern Oregon. 8.5" X 11", 310 ppgs. Retail Price: $29.99

Metal Mines of North West Oregon - Oregon mining historian Kerby Jackson introduces us to a classic work on Oregon's mining history in this important re-issue of Bulletin 14D, otherwise known as the North West Oregon Metal Mines Handbook. Unavailable since 1951, this important publication was originally compiled by the Oregon Department of Geology and Mineral Industries and includes detailed descriptions, histories and the geology of over 250 metallic mineral mines and prospects in North Western Oregon. 8.5" X 11", 182 ppgs. Retail Price: $19.99

Mines and Prospects of Oregon - Mining historian Kerby Jackson introduces us to a classic mining work by the Oregon Bureau of Mines in this important re-issue of The Handbook of Mines and Prospects of Oregon. Unavailable since 1916, this publication includes important insights into hundreds of gold, silver, copper, coal, limestone and other mines that operated in the State of Oregon around the turn of the 19th Century. Included are not only geological details on early mines throughout Oregon, but also insights into their history, production, locations and in some cases, also included are rare maps of their underground workings. 8.5" X 11", 314 ppgs. Retail Price: $24.99

Lode Gold of the Klamath Mountains of Northern California and South West Oregon (See California Mining Books)

Mineral Resources of South West Oregon - Unavailable since 1914, this publication includes important insights into dozens of mines that once operated in South West Oregon, including the famous gold fields of Josephine and Jackson Counties, as well as the Coal Mines of Coos County. Included are not only geological details on early mines throughout South West Oregon, but also insights into their history, production and locations. 8.5" X 11", 154 ppgs. Retail Price: $11.99

Chromite Mining in The Klamath Mountains of California and Oregon (See California Mining Books)

Southern Oregon Mineral Wealth - Unavailable since 1904, this rare publication provides a unique snapshot into the mines that were operating in the area at the time. Included are not only geological details on early mines throughout South West Oregon, but also insights into their history, production and locations. Some of the mining areas include Grave Creek, Greenback, Wolf Creek, Jump Off Joe Creek, Granite Hill, Galice, Mount Reuben, Gold Hill, Galls Creek, Kane Creek, Sardine Creek, Birdseye Creek, Evans Creek, Foots Creek, Jacksonville, Ashland, the Applegate River, Waldo, Kerby and the Illinois River, Althouse and Sucker Creek, as well as insights into local copper mining and other topics. 8.5" X 11", 64 ppgs. Retail Price: $8.99

Geology and Ore Deposits of the Takilma and Waldo Mining Districts - Unavailable since the 1933, this publication was originally compiled by the United States Geological Survey and includes details on gold and copper mining in the Takilma and Waldo Districts of Josephine County, Oregon. The Waldo and Takilma mining districts are most notable for the fact that the earliest large scale mining of placer gold and copper in Oregon took place in these two areas. Included in this report are details about some of the earliest large gold mines in the state such as the Llano de Oro, High Gravel, Cameron, Platerica, Deep Gravel and others, as well as copper mines such as the famous Queen of Bronze mine, the Waldo, Lily and Cowboy mines. In addition to geological examinations, insights are also provided into the production, day to day operations and early histories of these mines, as well as calculations of known mineral reserves in the area. This volume also includes six maps and 20 original illustrations. 8.5" X 11", 74 ppgs. Retail Price: $9.99

Gold Mines of Oregon - Oregon mining historian Kerby Jackson introduces us to a classic work on Oregon's mining history in this important re-issue of Bulletin 61, otherwise known as "Gold and Silver In Oregon". Unavailable since 1968, this important publication was originally compiled by geologists Howard C. Brooks and Len Ramp of the Oregon Department of Geology and Mineral Industries and includes detailed descriptions, histories and the geology of over 450 gold mines Oregon. Included are notes on the history, geology and gold production statistics of all the major mining areas in Oregon including the Klamath Mountains, the Blue Mountains and the North Cascades. While gold is where you find it, as every miner knows, the path to success is to prospect for gold where it was previously found. **8.5" X 11", 344 ppgs. Retail Price: $24.99**

Mines and Mineral Resources of Curry County Oregon - Originally published in 1916, this important publication on Oregon Mining has not been available for nearly a century. Included are rare insights into the history, production and locations of dozens of gold mines in Curry County, Oregon, as well as detailed information on important Oregon mining districts in that area such as those at Agness, Bald Face Creek, Mule Creek, Boulder Creek, China Diggings, Collier Creek, Elk River, Gold Beach, Rock Creek, Sixes River and elsewhere. Particular attention is especially paid to the famous beach gold deposits of this portion of the Oregon Coast. **8.5" X 11", 140 ppgs. Retail Price: $11.99**

Chromite Mining in South West Oregon - Originally published in 1961, this important publication on Oregon Mining has not been available for nearly a century. Included are rare insights into the history, production and locations of nearly 300 chromite mines in South Western Oregon. **8.5" X 11", 184 ppgs. Retail Price: $14.99**

Mineral Resources of Douglas County Oregon - Originally published in 1972, this important publication on Oregon Mining has not been available for nearly forty years. Included are rare insights into the geology, history, production and locations of numerous gold mines and other mining properties in Douglas County, Oregon. **8.5" X 11", 124 ppgs. Retail Price: $11.99**

Mineral Resources of Coos County Oregon - Originally published in 1972, this important publication on Oregon Mining has not been available for nearly forty years. Included are rare insights into the geology, history, production and locations of numerous gold mines and other mining properties in Coos County, Oregon. **8.5" X 11", 100 ppgs. Retail Price: $11.99**

Mineral Resources of Lane County Oregon - Originally published in 1938, this important publication on Oregon Mining has not been available for nearly seventy five years. Included are extremely rare insights into the geology and mines of Lane County, Oregon, in particular in the Bohemia, Blue River, Oakridge, Black Butte and Winberry Mining Districts. **8.5" X 11", 82 ppgs. Retail Price: $9.99**

Mineral Resources of the Upper Chetco River of Oregon: Including the Kalmiopsis Wilderness - Originally published in 1975, this important publication on Oregon Mining has not been available for nearly forty years. Withdrawn under the 1872 Mining Act since 1984, real insight into the minerals resources and mines of the Upper Chetco River has long been unavailable due to the remoteness of the area. Despite this, the decades of battle between property owners and environmental extremists over the last private mining inholding in the area has continued to pique the interest of those interested in mining and other forms of natural resource use. Gold mining began in the area in the 1850's and has a rich history in this geographic area, even if the facts surrounding it are little known. Included are twenty two rare photographs, as well as insights into the Becca and Morning Mine, the Emmly Mine (also known as Emily Camp), the Frazier Mine, the Golden Dream or Higgins Mine, Hustis Mine, Peck Mine and others. **8.5" X 11", 64 ppgs. Retail Price: $8.99**

Gold Dredging in Oregon - Originally published in 1939, this important publication on Oregon Mining has not been available for nearly seventy five years. Included are extremely rare insights into the history and day to day operations of the dragline and bucketline gold dredges that once worked the placer gold fields of South West and North East Oregon in decades gone by. Also included are details into the areas that were worked by gold dredges in Josephine, Jackson, Baker and Grant counties, as well as the economic factors that impacted this mining method. This volume also offers a unique look into the values of river bottom land in relation to both farming and mining, in how farm lands were mined, re-soiled and reclamated after the dredges worked them. Featured are hard to find maps of the gold dredge fields, as well as rare photographs from a bygone era. **8.5" X 11", 86 ppgs. Retail Price: $8.99**

Quick Silver Mining in Oregon - Originally published in 1963, this important publication on Oregon Mining has not been available for over fifty years. This publication includes details into the history and production of Elemental Mercury or Quicksilver in the State of Oregon. **8.5" X 11", 238 ppgs. Retail Price: $15.99**

Mines of the Greenhorn Mining District of Grant County Oregon - Originally published in 1948, this important publication on Oregon Mining has not been available for over sixty five years. In this publication are rare insights into the mines of the famous Greenhorn Mining District of Grant County, Oregon, especially the famous Morning Mine. Also included are details on the Tempest, Tiger, Bi-Metallic, Windsor, Psyche, Big Johnny, Snow Creek, Banzette and Paramount Mines, as well as prospects in the vicinities in the famous mining areas of Mormon Basin, Vinegar Basin and Desolation Creek. Included are hard to find mine maps and dozens of rare photographs from the bygone era of Grant County's rich mining history. **8.5" X 11", 72 ppgs. Retail Price: $9.99**

Geology of the Wallowa Mountains of Oregon: Part I (Volume 1) - Originally published in 1938, this important publication on Oregon Mining has not been available for nearly seventy five years. Included are details on the geology of this unique portion of North Eastern Oregon. This is the first part of a two book series on the area. Accompanying the text are rare photographs and historic maps.**8.5" X 11", 92 ppgs. Retail Price: $9.99**

Geology of the Wallowa Mountains of Oregon: Part II (Volume 2) - Originally published in 1938, this important publication on Oregon Mining has not been available for nearly seventy five years. Included are details on the geology of this unique portion of North Eastern Oregon. This is the first part of a two book series on the area. Accompanying the text are rare photographs and historic maps.**8.5" X 11", 94 ppgs. Retail Price: $9.99**

Field Identification of Minerals For Oregon Prospectors - Originally published in 1940, this important publication on Oregon Mining has not been available for nearly seventy five years. Included in this volume is an easy system for testing and identifying a wide range of minerals that might be found by prospectors, geologists and rockhounds in the State of Oregon, as well as in other locales. Topics include how to put together your own field testing kit and how to conduct rudimentary tests in the field. This volume is written in a clear and concise way to make it useful even for beginners. **8.5" X 11", 158 ppgs. Retail Price: $14.99**

The Bohemia Mining District of Oregon - Originally published in 1900, this important publication on Oregon Mining has not been available for over a century. Included in this volume are important insights into the famous Bohemia Mining District of Oregon, including the histories and locations of important gold mines in the area such as the Ophir Mine, Clarence, Acturas, Peek-a-boo, White Swan, Combination Mine, the Musick Mine, The California, White Ghost, The Mystery, Wall Street, Vesuvius, Story, Lizzie Bullock, Delta, Elsie Dora, Golden Slipper, Broadway, Champion Mine, Knott, Noonday, Helena, White Wings, Riverside and others. Also included are notes on the nearby Blue River Mining District. **8.5" X 11", 58 ppgs. Retail Price: $9.99**

The Gold Fields of Eastern Oregon - Unavailable since 1900, this publication was originally compiled by the Baker City Chamber of Commerce Offering important insights into the gold mining history of Eastern Oregon, "The Gold Fields of Eastern Oregon" sheds a rare light on many of the gold mines that were operating at the turn of the 19th Century in Baker County and Grant County in North Eastern Oregon. Some of the areas featured include the Cable Cove District, Baisely-Elhorn, Granite, Red Boy, Bonanza, Susanville, Sparta, Virtue, Vaughn, Sumpter, Burnt River, Rye Valley and other mining districts. Included is basic information on not only many gold mines that are well known to those interested in Eastern Oregon mining history, but also many mines and prospects which have been mostly lost to the passage of time. Accompanying are numerous rare photos **8.5" X 11", 78 ppgs. Retail Price: $10.99**

Gold Mining in Eastern Oregon - Originally published in 1938, this important publication on Oregon Mining has not been available for over a century. Included in this volume are important insights into the famous mining districts of Eastern Oregon during the late 1930's. Particular attention is given to those gold mines with milling and concentrating facilities in the Greenhorn, Red Boy, Alamo, Bonanza, Granite, Cable Cove, Cracker Creek, Virtue, Keating, Medical Springs, Sanger, Sparta, Chicken Creek, Mormon Basin, Connor Creek, Cornucopia and the Bull Run Mining Districts. Some of the mines featured include the Ben Harrison, North Pole-Columbia, Highland Maxwell, Baisley-Elkhorn, White Swan, Balm Creek, Twin Baby, Gem of Sparta, New Deal, Gleason, Gifford-Johnson, Cornucopia, Record, Bull Run, Orion and others. Of particular interest are the mill flow sheets and descriptions of milling operations of these mines. **8.5" X 11", 68 ppgs. Retail Price: $8.99**

The Gold Belt of the Blue Mountains of Oregon - Originally published in 1901, this important publication on Oregon Mining has not been available for over a century. Included in this volume are rare insights into the gold deposits of the Blue Mountains of North East Oregon, including the history of their early discovery and early production. Extensive details are offered on this important mining area's mineralogy and economic geology, as well as insights into nearby gold placers, silver deposits and copper deposits. Featured are the Elkhorn and Rock Creek mining districts, the Pocahontas district, Auburn and Minersville districts, Sumpter and Cracker Creek, Cable Cove, the Camp Carson district, Granite, Alamo, Greenhorn, Robinsonville, the Upper Burnt River Valley and Bonanza districts, Susanville, Quartzburg, Canyon Creek, Virtue, the Copper Butte district, the North Powder River, Sparta, Eagle Creek, Cornucopia, Pine Creek, Lower Powder River, the Upper Snake River Canyon, Rye Valley, Lower Burnt River Valley, Mormon Basin, the Malheur and Clarks Creek districts, Sutton Creek and others. Of particular interest are important details on numerous gold mines and prospects in these mining districts, including their locations, histories, geology and other important information, as well as information on silver, copper and fire opal deposits. **8.5" X 11", 250 ppgs. Retail Price: $24.99**

Mining in the Cascades Range of Oregon - Originally published in 1938, this important publication on Oregon Mining has not been available for over seventy five years. Included in this volume are rare insights into the gold mines and other types of metal mines in the Cascades Mountain Range of Oregon. Some of the important mining areas covered include the famous Bohemia Mining District, the North Santiam Mining District, Quartzville Mining District, Blue River Mining District, Fall Creek Mining District, Oakridge District, Zinc District, Buzzard-Al Sarena District, Grand Cove, Climax District and Barron Mining District. Of particular interest are important details on over 100 mines and prospects in these mining districts, including their locations, histories, geology and other important information. **8.5" X 11", 170 ppgs. Retail Price: $14.99**

Beach Gold Placers of the Oregon Coast - Originally published in 1934, this important publication on Oregon Mining has not been available for over 80 years. Included in this volume are rare insights into the beach gold deposits of the State of Oregon, including their locations, occurance, composition and geology. Of particular interest is information on placer platinum in Oregon's rich beach deposits. Also included are the locations and other information on some famous Oregon beach mines, including the Pioneer, Eagle, Chickamin, Iowa and beach placer mines north of the mouth of the Rogue River. **8.5" X 11", 60 ppgs. Retail Price: $8.99**

Idaho Mining Books

Gold in Idaho - Unavailable since the 1940's, this publication was originally compiled by the Idaho Bureau of Mines and includes details on gold mining in Idaho. Included is not only raw data on gold production in Idaho, but also valuable insight into where gold may be found in Idaho, as well as practical information on the gold bearing rocks and other geological features that will assist those looking for placer and lode gold in the State of Idaho. This volume also includes thirteen gold maps that greatly enhance the practical usability of the information contained in this small book detailing where to find gold in Idaho. **8.5" X 11", 72 ppgs. Retail Price: $9.99**

Geology of the Couer D'Alene Mining District of Idaho - Unavailable since 1961, this publication was originally compiled by the Idaho Bureau of Mines and Geology and includes details on the mining of gold, silver and other minerals in the famous Coeur D'Alene Mining District in Northern Idaho. Included are details on the early history of the Coeur D'Alene Mining District, local tectonic settings, ore deposit features, information on the mineral belts of the Osburn Fault, as well as detailed information on the famous Bunker Hill Mine, the Dayrock Mine, Galena Mine, Lucky Friday Mine and the infamous Sunshine Mine. This volume also includes sixteen hard to find maps. **8.5" X 11", 70 ppgs. Retail Price: $9.99**

The Gold Camps and Silver Cities of Idaho - Originally published in 1963, this important publication on Idaho Mining has not been available for nearly fifty years. Included are rare insights into the history of Idaho's Gold Rush, as well as the mad craze for silver in the Idaho Panhandle. Documented in fine detail are the early mining excitements at Boise Basin, at South Boise, in the Owyhees, at Deadwood, Long Valley, Stanley Basin and Robinson Bar, at Atlanta, on the famous Boise River, Volcano, Little Smokey, Banner, Boise Ridge, Hailey, Leesburg, Lemhi, Pearl, at South Mountain, Shoup and Ulysses, Yellow Jacket and Loon Creek. The story follows with the appearance of Chinese miners at the new mining camps on the Snake River, Black Pine, Yankee Fork, Bay Horse, Clayton, Heath, Seven Devils, Gibbonsville, Vienna and Sawtooth City. Also included are special sections on the Idaho Lead and Silver mines of the late 1800's, as well as the mining discoveries of the early 1900's that paved the way for Idaho's modern mining and mineral industry. Lavishly illustrated with rare historic photos, this volume provides a one of a kind documentary into Idaho's mining history that is sure to be enjoyed by not only modern miners and prospectors who still scour the hills in search of nature's treasures, but also those enjoy history and tromping through overgrown ghost towns and long abandoned mining camps. **8.5" X 11", 186 ppgs. Retail Price: $14.99**

Ore Deposits and Mining in North Western Custer County Idaho - Unavailable since 1913, this important publication was originally published by the Us Department of the Interior and has been unavailable for a century. Included are fine details on the geology, geography, gold placers and gold and silver bearing quartz veins of the mining region of North West Custer County, Idaho. Of particular interest is a rare look at the mines and prospects of the region, including those such as the Ramshorn Mine, SkyLark, Riverview, Excelsior, Beardsley, Pacific, Hoosier, Silver Brick, Forest Rose and dozens of others in the Bay Horse Mining District. Also covered are the mines of the Yankee Fork District such as the Lucky Boy, Badger, Black, Enterprise, Charles Dickens, Morrison, Golden Sunbeam, Montana, Golden Gate and others, as well as those in the Loon Mining District. **8.5" X 11", 126 ppgs. Retail Price: $12.99**

Gold Rush To Idaho - Unavailable since 1963, this important publication was originally published by the Idaho Bureau of Mines and has been unavailable for 50 years. "Gold Rush To Idaho" revisits the earliest years of the discovery of gold in Idaho Territory and introduces us to the conditions that the pioneer gold seekers met when they blazed a trail through the wilderness of Idaho's mountains and discovered the precious yellow metal at Oro Fino and Pierce. Subsequent rushes followed at places like Elk City, Newsome, Clearwater Station, Florence, Warrens and elsewhere. Of particular interest is a rare look at the hardships that the first miners in Idaho met with during their day to day existences and their attempts to bring law and order to their mining camps. 8.5" X 11", 88 ppgs. **Retail Price: $9.99**

The Geology and Mines of Northern Idaho and North Western Montana - Unavailable since 1909, this important publication was originally published by the Us Department of the Interior and has been unavailable for a century. Included are fine details on the geology and geography of the mining regions of Northern Idaho and North Western Montana. Of particular interest is a rare look at the mines and prospects of the region, including those in the Pine Creek Mining District, Lake Pend Oreille district, Troy Mining District, Sylvanite District, Cabinet Mining District, Prospect Mining District and the Missoula Valley. Some of the mines featured include the Iron Mountain, Silver Butte, Snowshoe, Grouse Mountain Mine and others. 8.5" X 11", 142 ppgs. **Retail Price: $12.99**

Mining in the Alturas Quadrangle of Blaine County Idaho - Unavailable since 1922, this important publication was originally published by the Idaho Bureau of Mines and has been unavailable for ninety years. Topics include the geology, rock formations and the formation of ore deposits in this important mining area of Idaho. Of particular focus is information on the local geology, quartz veins and ore deposits of this portion of Idaho. Included are hard to find details, including the descriptions and locations of numerous gold and silver mines in the area including the Silver King, Pilgrim, Columbia, Lone Jack, Sunbeam, Pride of the West, Lucky Boy, Scotia, Atlanta, Beaver-Bidwell and others mines and prospects. 8.5" X 11", 56 ppgs. **Retail Price: $8.99**

Mining in Lemhi County Idaho - Originally published in 1913, this important book on Idaho Mining has not been available to miners for over a century. Included are rare insights into hundreds of gold, silver, copper and other mines in this famous Idaho mining area. Details include the locations, geology, history, production and other facts of the mines of this region, not only gold and silver hardrock mines, but also gold placer mines, lead-silver deposits, copper mines, cobalt-nickel deposits, tungsten and tin mines . It is lavishly illustrated with hard to find photos of the period and rare mining maps. Some of the vicinities featured include the Nicholia Mining District, Spring Mountain District, Texas District, Blue Wing District, Junction District, McDevitt District, Pratt Creek, Eldorado District, Kirtley Creek, Carmen Creek, Gibbonsville, Indian Creek, Mineral Hill District, Mackinaw, Eureka District, Blackbird District, YellowJacket District, Gravel Range District, Junction District, Parker Mountain and other mining districts. 8.5" X 11", 226 ppgs. **Retail Price: $19.99**

Utah Mining Books

Fluorite in Utah - Unavailable since 1954, this publication was originally compiled by the USGS, State of Utah and U.S. Atomic Energy Commission and details the mining of fluorspar, also known as fluorite in the State of Utah. Included are details on the geology and history of fluorspar (fluorite) mining in Utah, including details on where this unique gem mineral may be found in the State of Utah. 8.5" X 11", 60 ppgs. **Retail Price: $8.99**

California Mining Books

The Tertiary Gravels of the Sierra Nevada of California - Mining historian Kerby Jackson introduces us to a classic mining work by Waldemar Lindgren in this important re-issue of The Tertiary Gravels of the Sierra Nevada of California. Unavailable since 1911, this publication includes details on the gold bearing ancient river channels of the famous Sierra Nevada region of California. 8.5" X 11", 282 ppgs. **Retail Price: $19.99**

The Mother Lode Mining Region of California - Unavailable since 1900, this publication includes details on the gold mines of California's famous Mother Lode gold mining area. Included are details on the geology, history and important gold mines of the region, as well as insights into historic mining methods, mine timbering, mining machinery, mining bell signals and other details on how these mines operated. Also included are insights into the gold mines of the California Mother Lode that were in operation during the first sixty years of California's mining history. 8.5" X 11", 176 ppgs. **Retail Price: $14.99**

Lode Gold of the Klamath Mountains of Northern California and South West Oregon - Unavailable since 1971, this publication was originally compiled by Preston E. Hotz and includes details on the lode mining districts of Oregon and California's Klamath Mountains. Included are details on the geology, history and important lode mines of the French Gulch, Deadwood, Whiskeytown, Shasta, Redding, Muletown, South Fork, Old Diggings, Dog Creek (Delta), Bully Choop (Indian Creek), Harrison Gulch, Hayfork, Minersville, Trinity Center, Canyon Creek, East Fork, New River, Denny, Liberty (Black Bear), Cecilville, Callahan, Yreka, Fort Jones and Happy Camp mining districts in California, as well as the Ashland, Rogue River, Applegate, Illinois River, Takilma, Greenback, Galice, Silver Peak, Myrtle Creek and Mule Creek districts of South Western Oregon. Also included are insights into the mineralization and other characteristics of this important mining region. 8.5" X 11", 100 ppgs. **Retail Price: $10.99**

Mines and Mineral Resources of Shasta County, Siskiyou County, Trinity County: California - Unavailable since 1915, this publication was originally compiled by the California State Mining Bureau and includes details on the gold mines of this area of Northern California. Also included are insights into the mineralization and other characteristics of this important mining region, as well as the location of historic gold mines. 8.5″ X 11″, 204 ppgs. **Retail Price: $19.99**

Geology of the Yreka Quadrangle, Siskiyou County, California - Unavailable since 1977, this publication was originally compiled by Preston E. Hotz and includes details on the geology of the Yreka Quadrangle of Siskiyou County, California. Also included are insights into the mineralization and other characteristics of this important mining region. 8.5″ X 11″, 78 **ppgs. Retail Price: $7.99**

Mines of San Diego and Imperial Counties, California - Originally published in 1914, this important publication on California Mining has not been available for a century. This publication includes important information on the early gold mines of San Diego and Imperial County, which were some of the first gold fields mined in California by early Spanish and Mexican miners before the 49ers came on the scene. Included are not only details on early mining methods in the area, production statistics and geological information, but also the location of the early gold mines that helped make California "The Golden State". Also included are details on the mining of other minerals such as silver, lead, zinc, manganese, tungsten, vanadium, asbestos, barite, borax, cement, clay, dolomite, fluospar, gem stones, graphite, marble, salines, petroleum, stronium, talc and others. 8.5″ X 11″, 116 **ppgs. Retail Price: $12.99**

Mines of Sierra County, California - Unavailable since 1920, this publication was originally compiled by the California State Mining Bureau and includes details on the gold mines of Sierra County, California. Also included are insights into the mineralization and other characteristics of this important mining region, as well as the location of historic gold mines. 8.5″ X 11″, 156 ppgs. **Retail Price: $19.99**

Mines of Plumas County, California - Unavailable since 1918, this publication was originally compiled by the California State Mining Bureau and includes details on the gold mines of Plumas County, California. Also included are insights into the mineralization and other characteristics of this important mining region, as well as the location of historic gold mines. 8.5″ X 11″, 200 ppgs. **Retail Price: $19.99**

Mines of El Dorado, Placer, Sacramento and Yuba Counties, California - Originally published in 1917, this important publication on California Mining has not been available for nearly a century. This publication includes important information on the early gold mines of El Dorado County, Placer County, Sacramento County and Yuba County, which were some of the first gold fields mined by the Forty-Niners during the California Gold Rush. Included are not only details on early mining methods in the area, production statistics and geological information, but also the location of the early gold mines that helped make California "The Golden State". Also included are insights into the early mining of chrome, copper and other minerals in this important mining area. 8.5″ X 11″, 204 ppgs. **Retail Price: $19.99**

Mines of Los Angeles, Orange and Riverside Counties, California - Originally published in 1917, this important publication on California Mining has not been available for nearly a century. This publication includes important information on the early gold mines of Los Angeles County, Orange County and Riverside County, which were some of the first gold fields mined in California by early Spanish and Mexican miners before the 49ers came on the scene. Included are not only details on early mining methods in the area, production statistics and geological information, but also the location of the early gold mines that helped make California "The Golden State". 8.5″ X 11″, 146 ppgs. **Retail Price: $12.99**

Mines of San Bernadino and Tulare Counties, California - Originally published in 1917, this important publication on California Mining has not been available for nearly a century. This publication includes important information on the early gold mines of San Bernadino and Tulare County, which were some of the first gold fields mined in California by early Spanish and Mexican miners before the 49ers came on the scene. Included are not only details on early mining methods in the area, production statistics and geological information, but also the location of the early gold mines that helped make California "The Golden State". Also included are details on the mining of other minerals such as copper, iron, lead, zinc, manganese, tungsten, vanadium, asbestos, barite, borax, cement, clay, dolomite, fluospar, gem stones, graphite, marble, salines, petroleum, stronium, talc and others. 8.5″ X 11″, 200 ppgs. **Retail Price: $19.99**

Chromite Mining in The Klamath Mountains of California and Oregon - Unavailable since 1919, this publication was originally compiled by J.S. Diller of the United States Department of Geological Survey and includes details on the chromite mines of this area of Northern California and Southern Oregon. Also included are insights into the mineralization and other characteristics of this important mining region, as well as the location of historic mines. Also included are insights into chromite mining in Eastern Oregon and Montana. 8.5″ X 11″, 98 ppgs. **Retail Price: $9.99**

Mines and Mining in Amador, Calaveras and Tuolumne Counties, California - Unavailable since 1915, this publication was originally compiled by William Tucker and includes details on the mines and mineral resources of this important California mining area. Included are details on the geology, history and important gold mines of the region, as well as insights into other local mineral resources such as asbestos, clay, copper, talc, limestone and others. Also included are insights into the mineralization and other characteristics of this important portion of California's Mother Lode mining region. 8.5" X 11", 198 ppgs. Retail Price: $14.99

The Cerro Gordo Mining District of Inyo County California - Unavailable since 1963, this publication was originally compiled by the United States Department of Interior. Included are insights into the mineralization and other characteristics of this important mining region of Southern California. Topics include the mining of gold and silver in this important mining district in Inyo County, California, including details on the history, production and locations of the Cerro Gordo Mine, the Morning Star Mine, Estelle Tunnel, Charles Lease Tunnel, Ignacio, Hart, Crosscut Tunnel, Sunset, Upper Newtown, Newtown, Ella, Perseverance, Newsboy, Belmont and other silver and gold mines in the Cerro Gordo Mining District. This volume also includes important insights into the fossil record, geologic formations, faults and other aspects of economic geology in this California mining district. 8.5" X 11", 104 ppgs. Retail Price: $10.99

Mining in Butte, Lassen, Modoc, Sutter and Tehama Counties of California - Unavailable since 1917, this publication was originally compiled by the United States Department of Interior. Included are insights into the mineralization and other characteristics of this important mining region of California. Topics include the mining of asbestos, chromite, gold, diamonds and manganese in Butte County, the mining of gold and copper in the Hayden Hill and Diamond Mountain mining districts of Lassen County, the mining of coal, salt, copper and gold in the High Grade and Winters mining districts of Modoc County, gold mining in Sutter County and the mining of gold, chromite, manganese and copper in Tehama County. This volume also includes the production records and locations of numerous mines in this important mining region. 8.5" X 11", 114 ppgs. Retail Price: $11.99

Mines of Trinity County California - Originally published in 1965, this important publication on California Mining has not been available for nearly fifty years. This publication includes important information on mines and mining in Trinity County, California, as well insights into the mineralization and geology of this important mining area in Northern California. Included are extensive details on hardrock and placer gold mines and prospects, including charts showing the locations of these historic mines.. 8.5" X 11", 144 ppgs. Retail Price: $12.99

Mines of Kern County California - Originally published in 1962, this important publication on California Mining has not been available for nearly fifty years. This publication includes important information on mines and mining in Kern County, California, as well insights into the mineralization and geology of this important mining area in California. Included are extensive details on hardrock and placer gold mines and prospects, including charts showing the locations of these historic mines. 8.5" X 11", 398 ppgs. Retail Price: $24.99

Mines of Calaveras County California - Originally published in 1962, this important publication on California Mining has not been available for nearly fifty years. This publication includes important information on mines and mining in Calaveras County, California, as well insights into the mineralization and geology of this important mining area in Northern California. Included are extensive details on hardrock and placer gold mines and prospects, including charts showing the locations of these historic mines. 8.5" X 11", 236 ppgs. Retail Price: $19.99

Lode Gold Mining in Grass Valley California - Unavailable since 1940, this publication was originally compiled by the United States Department of Interior. Included are insights into the gold mineralization and other characteristics of this important mining region of Nevada County, California. This volume also includes important insights into the geologic formations, faults and other aspects of economic geology in this California mining district. Of particular interest are the fine details on many hardrock gold mines in the area, including their locations, histories, development and mineralization. Some of the mines featured include the Gold Hill Mine, Massachusetts Hill, Boundary, Peabody, Golden Center, North Star, Omaha, Lone Jack, Homeward Bound, Hartery, Wisconsin, Allison Ranch, Phoenix, Kate Hayes, W.Y.O.D., Empire, Rich Hill, Daisy Hill, Orleans, Sultana, Centennial, Conlin, Ben Franklin, Crown Point and many others. 8.5" X 11", 148 ppgs. Retail Price: $12.99

Lode Mining in the Alleghany District of Sierra County California - Unavailable since 1913, this publication was originally compiled by the United States Department of Interior. Included are insights into the mineralization and other characteristics of this important mining region of Sierra County. Included are details on the history, production and locations of numerous hardrock gold mines in this famous California area, including the Tightner Mine, Minnie D., Osceola, Eldorado, Twenty One, Sherman, Kenton, Oriental, Rainbow, Plumbago, Irelan, Gold Canyon, North Fork, Federal, Kate Hardy and others. This volume also includes important insights into the fossil record, geologic formations, faults and other aspects of economic geology in this California mining district. 8.5" X 11", 48 ppgs. Retail Price: $7.99

Six Months In The Gold Mines During The California Gold Rush - Unavailable since 1850, this important work is a first hand account of one "49'ers" personal experience during the great California Gold Rush, shedding important light on one of the most exciting periods in the history of not only California, but also the world. Compiled from journals written between 1847 and 1849 by E. Gould Buffum, a native of New York, "Six Months In The Gold Mines During The California Gold Rush" offers a rare look into the day to day lives of the people who came to California to work in her gold mines when the state was still a great frontier. 8.5" X 11", 290 ppgs. **Retail Price: $19.99**

Quartz Mines of the Grass Valley Mining District of California - Unavailable since 1867, this important publication has not been available since those days. This rare publication offers a short dissertation on the early hardrock mines in this important mining district in the California Mother Lode region between the 1850's and 1860's. Also included are hard to find details on the mineralization and locations of these mines, as well as how they were operated in those day. 8.5" X 11", 44 ppgs. **Retail Price: $8.99**

Alaska Mining Books

Ore Deposits of the Willow Creek Mining District, Alaska - Unavailable since 1954, this hard to find publication includes valuable insights into the Willow Creek Mining District near Hatcher Pass in Alaska. The publication includes insights into the history, geology and locations of the well known mines in the area, including the Gold Cord, Independence, Fern, Mabel, Lonesome, Snowbird, Schroff-O'Neil, High Grade, Marion Twin, Thorpe, Webfoot, Kelly-Willow, Lane, Holland and others. 8.5" X 11", 96 ppgs. **Retail Price: $9.99**

The Juneau Gold Belt of Alaska - Unavailable since 1906, this hard to find publication includes valuable insights into the gold mines around Juneau, Alaska. The publication includes important details into the history, geology and locations of the well known gold mines and prospects in the area, including those around Windham Bay, Holkham Bay, Port Snettisham, on Grindstone and Rhine Creeks, Gold Creek, Douglas Island, Salmon Creek, Lemon Creek, Nugget Creek, from the Mendenhall River to Berners Bay, McGinnis Creek, Montana Creek, Peterson Creek, Windfall Creek, the Eagle River, Yankee Basin, Yankee Curve, Kowee Creek and elsewhere. Not only are gold placer mines included, but also hardrock gold mines. 8.5" X 11", 224 ppgs. **Retail Price: $19.99**

Arizona Mining Books

Mines and Mining in Northern Yuma County Arizona - Originally published in 1911, this important publication on Arizona Mining has not been available for over a hundred years. Included are rare insights into the gold, silver, copper and quicksilver mines of Yuma County, Arizona together with hard to find maps and photographs. Some of the mines and mining districts featured include the Planet Copper Mine, Mineral Hill, the Clara Consolidated Mine, Viati Mine, Copper Basin prospect, Bowman Mine, Quartz King, Billy Mack, Carnation, the Wardwell and Osbourne, Valensuella Copper, the Mariquita, Colonial Mine, the French American, the New York-Plomosa, Guadalupe, Lead Camp, Mudersbach Copper Camp, Yellow Bird, the Arizona Northern (Salome Strike), Bonanza (Harqua Hala), Golden Eagle, Hercules, Socorro and others. 8.5" X 11", 144 ppgs. **Retail Price: $11.99**

The Aravaipa and Stanley Mining Districts of Graham County Arizona - Originally published in 1925, this important publication on Arizona Mining has not been available for nearly ninety years. Included are rare insights into the gold and silver mines of these two important mining districts, together with hard to find maps. 8.5" X 11", 140 ppgs. **Retail Price: $11.99**

Gold in the Gold Basin and Lost Basin Mining Districts of Mohave County, Arizona - This volume contains rare insights into the geology and gold mineralization of the Gold Basin and Lost Basin Mining Districts of Mohave County, Arizona that will be of benefit to miners and prospectors. Also included is a significant body of information on the gold mines and prospects of this portion of Arizona. This volume is lavishly illustrated with rare photos and mining maps. 8.5" X 11", 188 ppgs. **Retail Price: $19.99**

Mines of the Jerome and Bradshaw Mountains of Arizona - This important publication on Arizona Mining has not been available for ninety years. This volume contains rare insights into the geology and ore deposits of the Jerome and Bradshaw Mountains of Arizona that will be of benefit to miners and prospectors who work those areas. Included is a significant body of information on the mines and prospects of the Verde, Black Hills, Cherry Creek, Prescott, Walker, Groom Creek, Hassayampa, Bigbug, Turkey Creek, Agua Fria, Black Canyon, Peck, Tiger, Pine Grove, Bradshaw, Tintop, Humbug and Castle Creek Mining Districts. This volume is lavishly illustrated with rare photos and mining maps. 8.5" X 11", 218 ppgs. **Retail Price: $19.99**

The Ajo Mining District of Pima County Arizona - This important publication on Arizona Mining has not been available for nearly seventy years. This volume contains rare insights into the geology and mineralization of the Ajo Mining District in Pima County, Arizona and in particular the famous New Cornelia Mine. 8.5" X 11", 126 ppgs. **Retail Price: $11.99**

Mining in the Santa Rita and Patagonia Mountains of Arizona - Originally published in 1915, this important publication on Arizona Mining has not been available for nearly a century. Included are rare insights into hundreds of gold, silver, copper and other mines in this famous Arizona mining area. Details include the locations, geology, history, production and other facts of the mines of this region. 8.5" X 11", 394 ppgs. **Retail Price: $24.99**

Mining in the Bisbee Quadrangle of Arizona - Originally published in 1906, this important publication on Arizona Mining has not been available for nearly a century. Included are rare insights into hundreds of gold, silver, copper and other mines in this famous Arizona mining area. Details include the locations, geology, history, production and other facts of the mines of this important mining region. 8.5" X 11", 188 ppgs. **Retail Price: $14.99**

Montana Mining Books

A History of Butte Montana: The World's Greatest Mining Camp - First published in 1900 by H.C. Freeman, this important publication sheds a bright light on one of the most important mining areas in the history of The West. Together with his insights, as well as rare photographs of the periods, Harry Freeman describes Butte and its vicinity from its early beginnings, right up to its flush years when copper flowed from its mines like a river. At the time of publication, Butte, Montana was known worldwide as "The Richest Mining Spot On Earth" and produced not only vast amounts of copper, but also silver, gold and other metals from its mines. Freeman illustrates, with great detail, the most important mines in the vicinity of Butte, providing rare details on their owners, their history and most importantly, how the mines operated and how their treasures were extracted. Of particular interest are the dozens of rare photographs that depict mines such as the famous Anaconda, the Silver Bow, the Smoke House, Moose, Paulin, Buffalo, Little Minah, the Mountain Consolidated, West Greyrock, Cora, the Green Mountain, Diamond, Bell, Parnell, the Neversweat, Nipper, Original and many others. 8.5" X 11", 142 ppgs. **Retail Price: $12.99**

The Butte Mining District of Montana - This important publication on Montana Mining has not been available for over a century. Included are rare insights into the gold, copper and silver mines of Butte, Montana together with hard to find maps and photographs. Some of the topics include the early history of gold, silver and copper mining in the Butte area, insight into the geology of its mining areas, the local distribution of gold, silver and copper ores, as well their composition and how to identify them. Also included are detailed facts about the mines in the Butte Mining District, including the famous Anaconda Mine, Gagnon, Parrot, Blue Vein, Moscow, Poulin, Stella, Buffalo, Green Mountain, Wake Up Jim, the Diamond-Bell Group, Mountain Consolidated, East Greyrock, West Greyrock, Snowball, Corra, Speculator, Adirondack, Miners Union, the Jessie-Edith May Group, Otisco, Iduna, Colorado, Lizzie, Cambers, Anderson, Hesperus, Preferencia and dozens of others. 8.5" X 11", 298 ppgs. **Retail Price: $24.99**

Mines of the Helena Mining Region of Montana - This important publication on Montana Mining has not been available for over a century. Included are rare insights into the gold, copper and silver mines of the vicinity of Helena, Montana, including the Marysville Mining District, Elliston Mining District, Rimini Mining District, Helena Mining District, Clancy Mining District, Wickes Mining District, Boulder and Basin Mining Districts and the Elkhorn Mining District. Some of the topics include the early history of gold, silver and copper mining in the Helena area, insight into the geology of its mining areas, the local distribution of gold, silver and copper ores, as well their composition and how to identify them. Also included are detailed facts, history, geology and locations of over one hundred gold, silver and copper mines in the area . 8.5" X 11", 162 ppgs, **Retail Price: $14.99**

Mines and Geology of the Garnet Range of Montana - This important publication on Montana Mining has not been available for over a century. Included are rare insights into the gold, copper and silver mines of the vicinity of this important mining area of Montana. Some of the topics include the early history of gold, silver and copper mining in the Garnet Mountains, insight into the geology of its mining areas, the local distribution of gold, silver and copper ores, as well their composition and how to identify them. Also included are detailed facts, history, geology and locations of numerous gold, silver and copper mines in the area . 8.5" X 11", 100 ppgs, **Retail Price: $11.99**

Mines and Geology of the Philipsburg Quadrangle of Montana - This important publication on Montana Mining has not been available for over a century. Included are rare insights into the gold, copper and silver mines of the vicinity of this important mining area of Montana. Some of the topics include the early history of gold, silver and copper mining in the Philipsburg Quadrangle, insight into the geology of its mining areas, the local distribution of gold, silver and copper ores, as well their composition and how to identify them. Also included are detailed facts, history, geology and locations of over one hundred gold, silver and copper mines in the area 8.5" X 11", 290 ppgs, **Retail Price: $24.99**

Geology of the Marysville Mining District of Montana - Included are rare insights into the mining geology of the Marysville Mining District. Some of the topics include the early history of gold, silver and copper mining in the area, insight into the geology of its mining areas, the local distribution of gold, silver and copper ores, as well their composition and how to identify them. Also included are detailed facts, history, geology and locations of gold, silver and copper mines in the area 8.5" X 11", 198 ppgs, **Retail Price: $19.99**

<u>The Geology and Mines of Northern Idaho and North Western Montana</u>

See listing under Idaho.

Nevada Mining Books

<u>The Bull Frog Mining District of Nevada</u> - Unavailable since 1910, this publication was originally compiled by the United States Department of Interior. This volume also includes important insights into the geologic formations, faults and other aspects of economic geology in this Nevada mining district. Of particular interest are the fine details on many mines in the area, including their locations, histories, development and mineralization. Some of the mines featured include the National Bank Mine, Providence, Gibraltor, Tramps, Denver, Original Bullfrog, Gold Bar, Mayflower, Homestake-King and other mines and prospects. **8.5" X 11", 152 ppgs, Retail Price: $14.99**

<u>History of the Comstock Lode</u> - Unavailable since 1876, this publication was originally released by John Wiley & Sons. This volume also includes important insights into the famous Comstock Lode of Nevada that represented the first major silver discovery in the United States. During its spectacular run, the Comstock produced over 192 million ounces of silver and 8.2 million ounces of gold. Not only did the Comstock result in one of the largest mining rushes in history and yield immense fortunes for its owners, but it made important contributions to the development of the State of Nevada, as well as neighboring California. Included here are important details on not only the early development and history of the Comstock, but also rare early insight into its mines, ore and its geology. **8.5" X 11", 244 ppgs, Retail Price: $19.99**

Colorado Mining Books

<u>Ores of The Leadville Mining District</u> - Unavailable since 1926, this publication was originally compiled by the United States Department of Interior. This volume also includes important insights into the ores and mineralization of the Leadville Mining District in Colorado. Topics include historic ore prospecting methods, local geology, insights into ore veins and stockworks, the local trend and distribution of ore channels, reverse faults, shattered rock above replacement ore bodies, mineral enrichment in oxidized and sulphide zones and more. **8.5" X 11", 66 ppgs, Retail Price: $8.99**

<u>Mining in Colorado</u> - Unavailable since 1926, this publication was originally compiled by the United States Department of Interior. This volume also includes important insights into the mining history of Colorado from its early beginnings in the 1850's right up to the mid 1920's. Not only is Colorado's gold mining heritage included, but also its silver, copper, lead and zinc mining industry. Each mining area is treated separately, detailing the development of Colorado's mines on a county by county basis. **8.5" X 11", 284 ppgs, Retail Price: $19.99**

<u>Gold Mining in Gilpin County Colorado</u> - Unavailable since 1876, this publication was originally compiled by the Register Steam Printing House of Central City, Colorado. A rare glimpse at the gold mining history and early mines of Gilpin County, Colorado from their first discovery in the 1850's up to the "flush years" of the mid 1870's. Of particular interest is the history of the discovery of gold in Gilpin County and details about the men who made those first strikes. Special focus is given to the early gold mines and first mining districts of the area, many of which are not detailed in other books on Colorado's gold mining history. **8.5" X 11", 156 ppgs, Retail Price: $12.99**

<u>Mining in the Gold Brick Mining District of Colorado</u> - Important insights into the history of the Gold Brick Mining District, as well as its local geography and economic geology. Also included are the histories and locations of historic mines in this important Colorado Mining District, including the Cortland, Carter, Raymond, Gold Links, Sacramento, Bassick, Sandy Hook, Chronicle, Grand Prize, Chloride, Granite Mountain, Lucille, Gray Mountain, Hilltop, Maggie Mitchell, Silver Islet, Revenue, Roosevelt, Carbonate King and others. In addition to hardrock mining, are also included are details on gold placer mining in this portion of Colorado. **8.5" X 11", 140 ppgs, Retail Price: $12.99**

Washington Mining Books

<u>The Republic Mining District of Washington</u> - Unavailable since 1910, this important publication was originally published by the Washington Geologic Survey and has been unavailable for a century. Topics include the geology, rock formations and the formation of ore deposits in this important mining area of Washington State. Also included are hard to find details on the geology, history and locations of dozens of mines in the area. Some of the mines featured include the New Republic Mine, Ben Hur, Morning Glory, the South Republic Mine, Quilp, Surprise, Black Tail, Lone Pine, San Poil, Mountain Lion, Tom Thumb, Elcaliph and many others. **8.5" X 11", 94 ppgs, Retail Price: $10.99**

The Myers Creek and Nighthawk Mining Districts of Washington - Unavailable since 1911, this important publication was originally published by the Washington Geologic Survey and has been unavailable for a century. Topics include the geology, rock formations and the formation of ore deposits in these important mining areas of Washington State. Also included are hard to find details on the geology, history and locations of dozens of mines in the area. Some of the mines featured include the Grant Mine, Monterey, Nip and Tuck, Myers Creek, Number Nine, Neutral, Rainbow, Aztec, Crystal Butte, Apex, Butcher Boy, Molson, Mad River, Olentangy, Delate, Kelsey, Golden Chariot, Okanogan, Ohio, Forty-Ninth Parallel, Nighthawk, Favorite, Little Chopaka, Summit, Number One, California, Peerless, Caaba, Prize Group, Ruby, Mountain Sheep, Golden Zone, Rich Bar, Similkameen, Kimberly, Triune, Hiawatha, Trinity, Hornsilver, Maquae, Bellevue, Bullfrog, Palmer Lake, Ivanhoe, Copper World and many others.
 8.5″ X 11″, 136 ppgs, Retail Price: $12.99

The Blewett Mining District of Washington - Unavailable since 1911, this important publication was originally published by the Washington Geologic Survey and has been unavailable for a century. Topics include the geology, rock formations and the formation of ore deposits in this important mining area of Washington State. Also included are hard to find details on the geology, history and locations of dozens of mines in the area. Some of the mines featured include the Washington Meteor, Alta Vista, Pole Pick, Blinn, North Star, Golden Eagle, Tip Top, Wilder, Golden Guinea, Lucky Queen, Blue Bell, Prospect, Homestake, Lone Rock, Johnson, and others. **8.5″ X 11″, 134 ppgs, Retail Price: $12.99**

Silver Mining In Washington - Unavailable since 1955, this important publication was originally published by the Washington Geologic Survey. Featured are the hard to find locations and details pertaining to Washington's silver mines. **8.5″ X 11″, 180 ppgs, Retail Price: $15.99**

The Mines of Snohomish County Washington - Unavailable since 1942, this important publication was originally published by the Washington Geologic Survey and has been unavailable for seventy years. Featured are details on a large number of gold, silver, copper, lead and other metallic mineral mines. Included are the locations of each historic mine, along with information on the commodity produced. **8.5″ X 11″, 98 ppgs, Retail Price: $10.99**

The Mines of Chelan County Washington - Unavailable since 1943, this important publication was originally published by the Washington Geologic Survey and has been unavailable for seventy years. Featured are details on a large number of gold, silver, copper, lead and other metallic mineral mines. Included are the locations of each historic mine, along with information on the commodity. **8.5″ X 11″, 88 ppgs, Retail Price: $9.99**

Metal Mines of Washington - Unavailable since 1921, this important publication was originally published by the Washington Geologic Survey and has been unavailable for nearly ninety years. Widely considered a masterpiece on the Washington Mining Industry, "Metal Mines of Washington" sheds light on the important details of Washington's early mining years. Featured are details on hundreds of gold, silver, copper, lead and other metallic mineral mines. Included are hard to find details on the mineral resources of this state, as well as the locations of historic mines. Lavishly illustrated with maps and historic photos and complete with a glossary to explain any technical terms found in the text, this is one of the most important works on mining in the State of Washington. No prospector or miner should be without it if they are interested in mining in Washington. **8.5″ X 11″, 396 ppgs, Retail Price: $24.99**

Gem Stones In Washington - Unavailable since 1949, this important publication was originally published by the Washington Geologic Survey and has been unavailable since first published. Included are details on where to find naturally occurring gem stones in the State of Washington, including quartz crystal, amethyst, smoky quartz, milky quartz, agates, bloodstone, carnelian, chert, flint, jasper, onyx, petrified wood, opal, fire opal, hyalite and others. **8.5″ X 11″, 54 ppgs, Retail Price: $8.99**

The Covada Mining District of Washington - Unavailable since 1913, this important publication was originally published by the Washington Geologic Survey and has been unavailable for a century. Topics include the geology, rock formations and the formation of ore deposits in this important mining area of Washington State. Also included are hard to find details on the geology, history and locations of dozens of mines in the area. Some of the mines featured include the Admiral, Advance, Algonkian, Big Bug, Big Chief, Big Joker, Black Hawk, Black Tail, Black Thorn, Captain, Cherokee Strip, Colorado, Dan Patch, Dead Shot, Etta, Good Ore, Greasy Run, Great Scott, Idora, IXL, Jay Bird, Kentucky Bell, King Solomon, Laurel, Laura S, Little Jay, Meteor, Neglected, Northern Light, Old Nell, Plymouth Rock, Polaris, Quandary, Reserve, Shoo Fly, Silver Plume, Three Pines, Vernie, White Rose and dozens of others. **8.5″ X 11″, 114 ppgs, Retail Price: $10.99**

The Index Mining District of Washington - Unavailable since 1912, this important publication was originally published by the Washington Geologic Survey and has been unavailable for a century. Topics include the geology, rock formations and the formation of ore deposits in this important mining area of Washington State. Also included are hard to find details on the geology, history and locations of dozens of mines in the area. Some of the mines featured include the Sunset, Non-Pareil, Ethel Consolidated, Kittaning, Merchant, Homestead, Co-operative, Lost Creek, Uncle Sam, Calumet, Florence-Rae, Bitter Creek, Index Peacock, Gunn Peak, Helena, North Star, Buckeye, Copper Bell, Red Cross and others. **8.5″ X 11″, 114 ppgs, Retail Price: $11.99**

Mining & Mineral Resources of Stevens County Washington - Unavailable since 1920, this important publication was originally published by the Washington Geologic Survey and has been unavailable for a century. Topics include the geology, rock formations and the formation of ore deposits in these important mining areas of Washington State. Also included are hard to find details on the geology, history and locations of hundreds of mines in the area. **8.5" X 11", 372 ppgs, Retail Price: $24.99**

The Mines and Geology of the Loomis Quadrangle Okanogan County, Washington - Unavailable since 1972, this important publication was originally published by the Washington Geologic Survey and has been unavailable for a century. Topics include the geology, rock formations and the formation of ore deposits in this important mining area of Washington State. Also included are hard to find details on the geology, history and locations of dozens of gold, copper, silver and other mines in the area. **8.5" X 11", 150 ppgs, Retail Price: $12.99**

The Conconully Mining District of Okanogan County Washington - Unavailable since 1973, this important publication was originally published by the Washington Geologic Survey and has been unavailable for a century. Topics include the geology, rock formations and the formation of ore deposits in this important mining area of Washington State, which also includes Salmon Creek, Blue Lake and Galena. Also included are hard to find details on the geology, mining history and locations of dozens of mines in the area. Some of the mines include Arlington, Fourth of July, Sonny Boy, First Thought, Last Chance, War Eagle-Peacock, Wheeler, Mohawk, Lone Star, Woo Loo Moo Loo, Keystone, Hughes, Plant-Callahan, Johnny Boy, Leuena, Gubser, John Arthur, Tough Nut, Homestake, Key and many others **8.5" X 11", 68 ppgs, Retail Price: $8.99**

Wyoming Mining Books

Mining in the Laramie Basin of Wyoming - Unavailable since 1909, this publication was originally compiled by the United States Department of Interior. Also included are insights into the mineralization and other characteristics of this important mining region, especially in regards to coal, limestone, gypsum, bentonite clay, cement, sand, clay and copper. **8.5" X 11", 104 ppgs, Retail Price: $11.99**

New Mexico Mining Books

The Mogollon Mining District of New Mexico - Unavailable since 1927, this important publication was originally published by the US Department of Interior and has been unavailable for 80 years. Topics include the geology, rock formations and the formation of ore deposits in this important mining area in New Mexico. Of particular focus is information on the history and production of the ore deposits in this area, their form and structure, vein filling, their paragenesis, origins and ore shoots, as well as oxidation and supergene enrichment. Also included are hard to find details, including the descriptions and locations of numerous gold, silver and other types of mines, including the Eureka, Pacific, South Alpine, Great Western, Enterprise, Buffalo, Mountain View, Floride, Gold Dust, Last Chance, Deadwood, Confidence, Maud S., Deep Down, Little Fanney, Trilby, Johnson, Alberta, Comet, Golden Eagle, Cooney, Queen, the Iron Crown, Eberle, Clifton, Andrew Jackson mine, Mascot and others. **8.5" X 11", 144 ppgs, Retail Price: $12.99**

The Percha Mining District of Kingston New Mexico - Unavailable since 1883, this important publication was originally published by the Kingston Tribune and has been unavailable for over one hundred and thirty five years. Having been written during the earliest years of gold and silver mining in the Percha Mining District, unlike other books on the subject, this work offers the unique perspective of having actually been written while the early mining history of this area was still being made. In fact, the work was written so early in the development of this area that many of the notable mines in the Percha District were less than a few years old and were still being operated by their original discoverers with the same enthusiasm as when they were first located. Included are hard to find details on the very earliest gold and silver mines of this important mining district near Kingston in Sierra County, New Mexico. **8.5" X 11", 68 ppgs, Retail Price: $9.99**

East Coast Mining Books

The Gold Fields of the Southern Appalachians - Unavailable since 1895, this important publication was originally published by the US Department of Interior and has been unavailable for nearly 120 years. Topics include the geology, rock formations and the formation of ore deposits in this important mining area of the American South. Of particular focus is information on the history and statistics of the ore deposits in this area, their form and structure and veins. Also included are details on the placer gold deposits of the region. The gold fields of the Georgian Belt, Carolinian Belt and the South Mountain Mining District of North Carolina are all treated in descriptive detail. Included are hard to find details, including the descriptions and locations of numerous gold mines in Georgia, North Carolina and elsewhere in the American South. Also included are details on the gold belts of the British Maritime Provinces and the Green Mountains. **8.5" X 11", 104 ppgs, Retail Price: $9.99**